SpringerBriefs in Molecular Science

History of Chemistry

Series editor

Seth C. Rasmussen, Department of Chemistry and Biochemistry,
North Dakota State University, Fargo, ND, USA

W0235040

For further volumes:
http://www.springer.com/series/10127

Gary Patterson

Polymer Science from 1935–1953

Consolidating the Paradigm

 Springer

Gary Patterson
Department of Chemistry
Carnegie Mellon University
Pittsburgh, PA
USA

ISSN 2212-991X
ISBN 978-3-662-43535-9   ISBN 978-3-662-43536-6   (eBook)
DOI 10.1007/978-3-662-43536-6
Springer Heidelberg New York Dordrecht London

Library of Congress Control Number: 2014940064

Printed on acid-free paper

Springer is part of Springer Science+Business Media (www.springer.com)

Acknowledgements

This book has benefitted from the assistance of many archives and many individual people. The story of Paul Flory has been developed in collaboration with James Mark of the University of Cincinnati. Information on Geoffrey Gee was provided by the University of Manchester Library Archives. Material on Flory's work at Goodyear Tire and Rubber Company was obtained from the Company Archives at the University of Akron. Information about Thomas G Fox was gleaned from the Carnegie Mellon University archives.

Special thanks goes to John Wiley and Sons publishers for permission to reproduce so many figures from old Interscience volumes. Also, Ursula Eirich Moeller graciously provided both her own family monograph on Fritz Eirich and a great picture. The Rubber Science Hall of Fame at the University of Akron was very helpful with both biographical materials and good pictures. The oral history program at the Chemical Heritage Foundation was the source of many biographical details. Peter J.T. Morris was a constant encouragement during the writing of this book. Guy Berry also provided personal reflections and keen insights. Since I knew many of the figures profiled in Chapter 5, I thank them for inspiration and examples.

Careful editing and good suggestions were provided by Seth Rasmussen of the University of North Dakota. And constant encouragement and great patience were shown by Elizabeth Hawkins of Springer.

Contents

Figures

Abstract

As documented in *A Prehistory of Polymer Science,* the scientific community devoted to macromolecules came of age in 1935 at the Faraday Society Discussion on Polymerization. The rapid progress of this field is presented in the present volume. The story is organized around several key figures: Paul Flory, Herman Mark, and Speed Marvel. Each one created a network of friends and colleagues that were very productive. Dozens of polymer scientists are noted in these three chapters. In addition, 12 people are discussed in more depth: Emil Ott, Maurice Huggins, Raymond Fuoss, L. R. G. Treloar, G. Stafford Whitby, Roelof Houwink, Walter Stockmayer, John Ferry, Bruno Zimm, Paul Doty, Dick Stein, and William O. Baker. During this time period, many key concepts were developed and remain central to the field today. The Gaussian chain model helps to unify the discussion of both equilibrium and dynamic properties of polymers. The understanding of polymer solutions as a function of concentration was greatly advanced by Flory and others. And the structure and dynamics of bulk polymers was explained. The era ends with the awarding of the Nobel Prize in Chemistry to Hermann Staudinger, but by then he was only a memory within the polymer community.

Keywords Paul Flory · Herman Mark · Carl Marvel · Gaussian chain · Excluded volume · Rubber · Polymer solutions · Molecular weight distributions · Intrinsic viscosity · Herman Staudinger

Chapter 1
Polymer Science 1935-1953: Consolidating the Paradigm

Introduction

1935 was a very good year for polymer science. More than 200 scientists, engineers and technologists had gathered in Manchester to present and discuss the field of polymerization and the discipline of polymer science. Outstanding leaders in both synthetic and physical chemistry had been identified and their work had solidified into a consilience of concepts and observations [1].

1937 was a very bad year for polymer science. The leading synthetic polymer chemist from DuPont, Wallace Hume Carothers (1896-1937), succumbed to a lifelong battle with depression and committed suicide. Leon Trotsky failed to overthrow Joseph Stalin, and Adolph Hilter set in motion the program of expansion and extermination known as World War II. Like the destruction of Jerusalem in A.D. 70, the Anschluss dispersed the intellectual leaders of European polymer science. Some fled to Switzerland and stayed; some fled to England and stayed; and some eventually came to the United States and stayed. Hermann Staudinger (1881-1965) remained in Germany and in 1940 founded the Institute for Macromolecular Chemistry at the University of Freiburg.

While the war years were a major disruption of scientific life, they also provided opportunities to pursue the polymer science of synthetic rubber. Both money and manpower were invested in this element of national security. Industrial firms were also quick to capitalize on the new knowledge of polymer science to introduce new polymers or to commercialize known materials [2].

Amidst the chaos of war, more quiet streams of scientific advance were also flowing. Universities in Europe, England and the United States were centers of intellectual activity that allowed scientists to think about the new phenomena of polymer science and apply the latest theoretical ideas to their explication. One of the fruits of this endeavor was the publication of "Principles of Polymer Chemistry" by Paul J. Flory (1910-1985) in 1953 [3]. With the awarding of the Nobel Prize in Chemistry to Hermann Staudinger in 1953 "for his discoveries in the field of macromolecular chemistry", the field of polymer science was firmly established as a thriving branch of intellectual and industrial activity. The present

G. Patterson, *Polymer Science from 1935–1953*, SpringerBriefs in History of Chemistry, DOI: 10.1007/978-3-662-43536-6_1, © The Author(s) 2014

volume surveys this period in the history of polymer science as the concepts proposed in 1935 were examined and extended. By 1953, most of the key concepts that define polymer science as a paradigm community were established.

The story can be organized around several key individuals and their friends and colleagues. Chapter 2 begins with the story of Paul Flory and his friends from DuPont to Cornell. The tale begins with Wallace Carothers and his seminal work in polymer science. It follows Flory as he moves to the University of Cincinnati and then to the Esso Laboratories of the Standard Oil Development Company of New Jersey. His collaboration with John Rehner, Jr. was both scientifically and personally fulfilling. The Goodyear Tire and Rubber Company took advantage of the newly funded Rubber Reserve Program to bring Flory to Akron as Head of a department devoted to fundamental polymer science. In this capacity Paul Flory forged lifelong bonds with Speed Marvel, Herman Mark, Fred Wall and Peter Debye. His best friend within Goodyear, Thomas G Fox, remained close to Flory until his own untimely death. After the war, Flory was called to Cornell where he flourished. Two of his postdoctoral fellows became close friends, William Krigbaum and Leo Mandelkern. The full story of Paul John Flory and his friends will appear in a forthcoming biography [4].

Chapter 3 is organized around Herman Mark (1895-1992). He was lovingly referred to as the *Geheimrat* and counted all polymer scientists as his friends. Once he became established at the Brooklyn Polytechnic Institute, this became the scientific capital of world polymer science. His close collaborator and alter-ego, Kurt H. Meyer, landed at the University of Geneva. Although the war did hinder communication, Mark published an English version of his monograph with Meyer as two volumes. Another compatriot of Mark that retreated to Switzerland was Werner Kuhn who took a physical chemistry post at the University of Basel. Mark was very interested in polymer reactions as well, and soon enlisted his former colleague from Vienna, R. Raff, to help him write an Interscience volume on this subject. Raff had a long career in Canada. One of Mark's greatest students was Turner Alfrey Jr.. He was also enlisted to produce an Interscience volume on the mechanical properties of polymers.

Chapter 4 is centered on Carl (Speed) Marvel (1894-1988). With the death of Wallace Carothers, Marvel took up the mantle as the American prophet of synthetic polymer chemistry. He practiced his religion at the University of Illinois, in the sunshine of Roger Adams. When I met him in 1972, he was still going strong at the University of Arizona. Only death led to his retirement. During the war he was a stalwart member of the Rubber Boys Club. After the war, he traveled throughout Europe as a representative of the United State government, with a mission to glean as much knowledge as he could. He did not disappoint. But, he also spread the gospel of polymer science and counted as one of his converts Karl Zeigler. This activity probably cost him any chance of a Nobel prize. One of his American converts was Fred Wall of the University of Illinois.

Chapter 5 includes biographies of twelve other outstanding figures in polymer science that achieved prominence in the period 1935-1953. They include: Emil Ott,

Maurice Huggins, Raymond Fuoss, Leslie Treloar, G. Stafford Whitby, Roelof Houwink, Walter Stockmayer, John Ferry, Bruno Zimm, Paul Doty, Richard Stein and William O. Baker. The number of stories that can be told about this period in the development of polymer science is very much larger than the page limit for this book.

Chapter 6 reviews the career of Hermann Staudinger and the awarding of the Nobel Prize in Chemistry. While Staudinger continued to carry out a large amount of research after 1935, he had lost his position as the leading proponent of the macromolecular paradigm. He was isolated inside Germany during the war, and isolated outside Germany after the war. His Institute was bombed. The awarding of the Nobel Prize delighted him and he and his beloved wife, Magda, wrote a fanciful hagiography published in 1965.

Chapter 7 includes some concluding remarks and reflections for the future. Both a look back and a look forward are presented.

Another organizing principle for the story is the key books that documented the establishment of paradigms in polymer science. Interscience published a series of books called "High Polymers". It was edited by Herman Mark and a constantly changing cast of other leaders in polymer science. Many of these volumes are reviewed in detail in this story. There is also an eclectic group of monographs that are still considered classics, such as L.R.G. Treloar's "The Physics of Rubber Elasticity" [5]. Any scientific community is partially defined by its favorite books, and the author of the present volume is the Chief Bibliophile of the Bolton Society of the Chemical Heritage Foundation.

The final organizing principle is the paradigms themselves. While the fundamental paradigm of the existence of covalent macromolecules underlies the whole field, and Staudinger was given the Nobel Prize for promoting it, there are dozens of key concepts that were developed and accepted in the period 1935-1953. Two of the key figures in this theoretical development were Werner Kuhn (1899-1963) and Paul Flory. They both used the Gaussian chain model to formulate the properties of macromolecules in dilute solution, concentrated solution, bulk polymer, and as either bulk rubbers or gels. One of the most elegant applications of this model was to the theory of light scattering in polymer solutions. The fundamental work of Peter Debye (1884-1966) was extended and completed by Flory and collaborators. Full experimental and theoretical verification of this work did not appear until the first decade of the 21st century [6].

The present volume follows the initial study: "A Prehistory of Polymer Science" [1]. The origins and development of the field were surveyed up to the time of the Faraday Discussion of 1935 on "Polymerization." From early studies of caoutchouc at the French Academy of Sciences to seminal syntheses of polyamides at DuPont, the concept of covalent macromolecules was developed. This fruitful concept was extended to a large number of observable properties in the period 1935-1953.

References

1. Patterson G (2012) A prehistory of polymer science. Springer, New York
2. Morris PJT (1989) The American synthetic rubber research program. University of Pennsylvania Press, Philadelphia
3. Flory PJ (1953) Principles of polymer chemistry. Cornell University Press, Ithaca
4. Fried J, Mark J, Patterson G (2015) Paul John Flory: a life of science and friends. CRC Press, Boca Raton
5. Treloar LRG (1949) The physics of rubber elasticity. Oxford at the Clarendon Press, Oxford
6. Francis RS, Patterson GD, Kim SH (2006) Liquid like structure of polymer solutions near the overlap concentration. J Pol Sci Part B Polym Phys 44:703–710

Chapter 2
Flory and Friends

High Polymers: Volume 1 "Collected Papers of Wallace Hume Carothers on High Polymeric Substances"

Although Wallace Carothers (1896-1937) was not alive to personally interact with the ongoing polymer science community, he left a legacy of publication that was quickly collected and edited by Herman Mark (1895-1992) and G. Stafford Whitby (1887-1972). It was issued in 1940 as Volume 1 in the Interscience series: "High Polymers" [1]. A short biography condensed from the official National Academy of Sciences version by Roger Adams (1889-1971) is included. The education of Wallace Carothers was greatly assisted by his time at the University of Illinois. In addition to research with Adams, Carothers was retained as an instructor until 1926. He was acclaimed as "one of the most brilliant students who had ever been awarded the doctor's degree at Illinois." He was then appointed to the Chemistry Department at Harvard University. James B. Conant (1893-1978), Professor of Organic Chemistry and later President of Harvard, rated Carothers very highly for his originality. While Carothers would go on to become famous for polymer chemistry, he was at the top of his field, both experimentally and theoretically, in synthetic and physical organic chemistry. It was his thorough grounding in the fundamentals of organic chemistry that enabled him to make rapid progress in his understanding of polymerization

Developments at DuPont opened new avenues in research for Carothers. He was hired in 1928 to direct a new laboratory of organic chemistry at the Experimental Station in Wilmington, Delaware. Dr. Elmer K. Bolton (1886-1968), Chemical Director of DuPont, was "impressed by the breadth and depth of his knowledge." Carothers wasted no time in pursuing the field of experimental synthetic polymer chemistry. His first paper on polymerization appeared in the *Journal of the American Chemical Society (JACS)* in 1929 [2]. It was already a masterpiece of clear thinking and encyclopedic presentation. Whereas

G. Patterson, *Polymer Science from 1935–1953*, SpringerBriefs in History of Chemistry, DOI: 10.1007/978-3-662-43536-6_2, © The Author(s) 2014

Staudinger was always rather confused in exposition, Carothers was both precise and elegant. A flurry of experimental work was summarized in the monumental paper entitled "Polymerization" published in 1931 in *Chemical Reviews* [3]. The basic principles of condensation polymerization were established beyond dispute. The whole landscape of polymeric substances was considered in detail. Any coupling reaction that was known to chemistry could be used in the synthesis of a polymer.

Carothers also extensively studied polymers formed by radical reactions involving unsaturated monomers. He made special contributions to systems involving both double and triple bonds. One of his earliest discoveries (1931) yielded the important rubber, polychloroprene [4]. He systematically investigated the influence of substituents on the reactivity of butadienes. Carothers carried out polymerizations of vinylacetylene and divinylacetyalene.

Chloroprene polymerizes thermally to yield a true rubbery solid with a characteristic liquid-like structure. Stretching produces crystallinity and the expected X-ray pattern for an oriented fiber. A detailed mechanistic analysis yielded the multiple structures that appear during the reaction. Another synthetic pathway leads to polychloroprene latex. Characterization of these particles with the ultra-centrifuge yielded a highly peaked distribution with a mean radius near 0.06 micron. Because of the very small particle size compared with natural rubber latex, this material can be used for many applications requiring minute particles. The detailed analysis and keen structural insight set a standard that is rarely met, even today.

The final paper in "The Collected Papers of Wallace Carothers on Polymerization" is the classic plenary lecture at the Faraday Discussion of 1935 reviewed in "A Prehistory of Polymer Science" [5]. Just as Humphry Davy's greatest gift to Chemistry was Michael Faraday, Wallace Carothers revealed in this paper that there was a new scientist working on fundamental properties of polymers: Paul John Flory (1910-1985).

Paul John Flory (1910-1985)

Flory was also from the Midwest and attended Ohio State University (Ph.D. 1934). His class notebooks reveal a lively and inquiring mind. He wrote his thesis on the photochemical decomposition of nitric oxide. A thorough grounding in chemical kinetics was the perfect preparation for his work with Carothers at DuPont. It was known that condensation polymerization produced a broad distribution of molecular weights. Flory was able to calculate from first principles the distribution of chain lengths for a system of reacting molecules as a function of extent of reaction, p. This paper appeared in *JACS* in 1936 [6]. The "most probable distribution law" for the distribution of x-mers is one of the foundations of polymer science: $\Pi_x = xp^{x-1}(1-p)^2$, where x is the number of monomer units in the polymer.

Once the concept of a distribution of chain lengths is accepted, the need to take into account the effect of this distribution on measured properties becomes apparent. Flory derived the number average distribution, the weight-average distribution and the z-average distribution. The number average distribution of chain lengths is the usual linear distribution where each chain length is weighted by the number fraction of chains in the total sample. Some physical properties depend on the weight fraction associated with each chain length. The z-fraction is the next higher weighting that depends on number fraction times the square of the chain length. From these distributions he calculated the number average molecular weight, the weight average molecular weight and the z-average molecular weight. Experimental properties such as the intrinsic viscosity depend on higher molecular weight averages. Staudinger had made limited progress in understanding the interpretation of the viscosity; it was time for leaders that understood physical chemistry.

Werner Kuhn (1899-1963)

Werner Kuhn (1899-1963) brought the power of theoretical chemical physics to the study of polymers. Kuhn was educated at the Eidgenossiche Technische Hochschule (ETH) in Zurich, Switzerland and received his doctorate in physical chemistry at the University of Zurich in 1923. He worked in many of the best laboratories in Europe, but with the approach of the War, he retreated to the University of Basel's Institute of Physical Chemistry in 1939. Kuhn was thoroughly familiar with the theory of molecules that developed in Europe in the 1930s. On this basis he proposed in 1936 that chain molecules in solution can be modeled as a random coil [7]. The chain molecule is represented in terms of vectors associated with each main chain bond. This model assumes that the direction of the main chain bonds eventually becomes uncorrelated when the distance along the chain becomes large. Kuhn worked out the mathematical consequences of this model in a form that is still used today. He was then able to use the random coil model to explain the source of the retractive force in rubber elasticity. The random coil model also allowed the problem of polymer solution viscosity to be attacked. Staudinger was just wrong on this subject. It took a real theorist to develop a real theory.

Polymers formed from unsaturated monomers also lead to distributions of molecular weight. But, the mechanism of the radical reaction was not yet well understood. Again, Flory's thorough background in chemical kinetics provided the insight to realize that individual growing chains could terminate by chain transfer to monomer or other chains. The classic paper on "The Mechanism of Vinyl Polymerization" appeared in *JACS* in 1937 [8]. The mechanism of vinyl polymerization had been considered by Staudinger [9], and by Dostal and Mark [10], but their theories did not agree with the experimental facts. The rate of the reaction increases with time as the concentration of radical centers increases. The average molecular weight of the formed polymers is relatively constant over much of the reaction. While the termination reactions due to radical coupling or disproportionation were well known for small molecules, these processes removed radicals from the reaction and would not lead to an increase in the rate. It was the proposal that individual chains could terminate by chain transfer that solved the problem. The radical center was merely transferred and remained active for further polymerization on another center. Paul Flory was not a one hit wonder!

With the death of Carothers, the character of the research laboratory at DuPont changed [11]. There was now no one devoted to fundamental research for its own sake, even in the pursuit of industrially relevant knowledge. Flory chose to leave DuPont and joined the Basic Sciences Research Laboratory of the University of Cincinnati in 1938. He continued his studies of the mechanisms and chain length distributions of polymers.

Another class of polymers grows by sequential addition of monomers without possibility of termination. The classic paper by Flory on the "Molecular Size Distribution in Ethylene Oxide Polymers" appeared in *JACS* in 1940 [12]. These polymers are observed to have a very narrow distribution of chain lengths. Flory was able to use statistical arguments in connection with the correct mechanism for these reactions to derive the Poisson distribution for the chain length:

$$\wp(x) = e^{-v} v^{x-1} / (x - 1)!$$

where v is the number average chain length less 1. The paradigm of molecular weight distributions was now firmly established and remains an essential part of every discussion of polymeric systems.

Another polymeric phenomenon that was attacked by Flory while at Cincinnati was the structure and formation of gels [13]. It had been known for many years that when a trifunctional or higher monomer was polymerized with a difunctional monomer, the system eventually became a three dimensional soft solid. At a well-defined point during the reaction the viscosity of the material diverged to infinity and the shear modulus started to rise above zero. Flory used statistical arguments and a keen understanding of the mechanism of the gelation reaction to derive relationships between the extent of reaction, the actual stoichiometry of the polymer network and the critical probability of infinite network formation. For a branching unit of functionality f, the critical value of the probability is: $\alpha_C = 1/(f - 1)$. At the gel point, the system still contains mostly unreacted monomer and a distribution of smaller polymers. The "sol" fraction then decreases as the reaction proceeds. A detailed calculation of the chain length

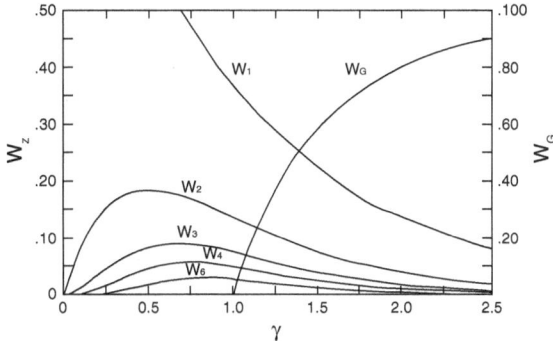

FIG. 3. Weight fractions (W_z, left ordinate scale) of species composed of one, two, three, four, and six chains *versus* cross-linking index γ. Weight fraction of ge, W_g (right ordinate scale).

Fig. 2.1 Gelation diagram from Flory [14] (J. Phys. Chem. by permission)

distribution in the sol followed in a second paper (See Fig. 2.1). And the calculations were extended to tetrafunctional junctions in a third contribution. This work was presented and summarized in a classic article in the *Journal of Physical Chemistry* in 1942 [14]. By this time Flory had joined the Esso Laboratories of the Standard Oil Development Company in Elizabeth, New Jersey. He was free to pursue a wide range of problems in this position. And he did!

Esso Laboratories

One of the other extensive areas that Flory pursued at Esso was the thermodynamics of polymer solutions. The general theory of two component liquid mixtures was still in its infancy in 1940. A series of papers by Flory published in the *Journal of Chemical Physics* changed all that. The first major contribution (1941) was the derivation of an expression for the entropy of mixing that included the **volume fraction** rather than the mole fraction of the solvent [15]. Standard physical chemistry texts still fail to acknowledge this feature, but all industrial laboratories use the correct Flory expression to obtain osmotic pressures that agree with experiment! A much more extensive paper that included the full discussion of the so-called Flory-Huggins theory appeared in 1942 [16]. It included a calculation of the two component phase diagram for partially miscible liquids that was at least qualitatively correct. Flory discusses in detail the limitations of the theory: random mixing is assumed, the volume change on mixing is ignored.

Geoffrey Gee (1910-1996)

Reference is made in this paper to another emerging leader, Geoffrey Gee (1910-1996, FRS). Gee was educated at the University of Manchester (M.Sc. 1933) and at Cambridge University (Ph.D. 1936) with Sir Eric Rideal (1890-1974, FRS). His work was much appreciated and he stayed on for a postdoctoral fellowship until 1938. He became an expert in the chemical kinetics of thin films. These were the glory years in the Rideal Laboratory at Cambridge. Sir Harry Melville (1908-2000, FRS, KCB) was also with Rideal at this point. Gee then joined the British Rubber Producers Research Association (BRPRA) as a Senior Physical Chemist. He carried out a program of fundamental research on the structure and properties of natural rubber. Another leading figure in the study of the physical properties of rubber at the BRPRA was L.R.G. Treloar (1906-1985). His monograph "The Physics of Rubber Elasticity" is still considered an essential volume [17]. Consideration of the osmotic pressure of rubber solutions and the swelling of crosslinked rubber led Gee to make some of the best thermodynamic measurements then published. They impressed Flory and the two men went on to become great friends. Gee could approach Flory as an equal and they were scientifically good for one another. It took more than simplified theory to create a sound foundation for polymer science. Flory always put more faith in the experiments than the theory, but he did try to understand polymer solutions and swollen rubber in the best quantitative way he could. He never stopped creating better theories of polymer solutions and gels. One of the foundational papers in polymer science is Gee's review in the initial volume of the *Quarterly Reviews of the Chemical Society* (1947) on "Some thermodynamic properties of high polymers and their molecular interpretation" [18]. Gee's Royal Society biographical memoir was expertly written by one of his most famous colleagues, Sir Geoffrey Allen (1928-, FRS) (Fig. 2.2).

One of the most pressing problems in polymer science in the early 1940s was the quantitative theory of rubber elasticity. Substantial conceptual progress had been made by Kuhn and by Eugene Guth (1905-1990) and Herman Mark (1895-1992), but a readily applicable expression was needed to guide further work. Flory collaborated with John Rehner Jr. (1908-1997) on a classic series of papers on polymer networks and gels. The random coil model was now accepted as a good part of the paradigm. Flory considered in detail the role of the network in the theory. The distribution of crosslink sites is deformed by the macroscopic deformation; Flory assumed an affine deformation. The distribution of end-to-end vectors for the elastically effective chains is modified by the network and the detailed equations are given. The new insights continued the development of the theory of rubber elasticity [19]. It is still not "finished."

With both a workable theory for network deformation, and a semi-quantitative theory of polymer solutions, Flory and Rehner tackled the problem of swollen rubber [20]. Their combined theory is so good, that it is still used in industry, even though there are serious problems at the most fundamental level. They even correctly described the change in the elongational modulus due to swelling. It is clear that Esso Laboratories provided a scientifically productive work environment.

Fig. 2.2 Sir Geoffrey Gee (FRS) (University of Manchester, by permission)

Goodyear Tire and Rubber Company

The large body of work produced at Esso vaulted Flory into a position of intellectual leadership in the polymer community. In 1943 Goodyear Tire and Rubber Company offered Flory an administrative leadership position as Head of its fundamental polymer research group in Akron, Ohio. It was a mutually beneficial situation.

Early in 1944 a symposium on the "Theory of Long-range Elasticity" was held at the Cleveland Meeting of the American Chemical Society. Flory's plenary lecture, "Network Structure and the Elastic Properties of Vulcanized Rubber", was published in *Chemical Reviews* [21]. It summarized all the work done through 1944 and presented new comparisons between theory and experiment. The classic case of the elongational modulus and its dependence on swelling is shown in Figure 8 from that paper (Fig. 2.3). It is based on the Flory-Rehner equation:

$$\tau \cong RT\left(\alpha - 1/\alpha^2\right)\left((1/2) - \chi\right)/\varphi_2^{5/3}\,\bar{V}_1$$

where α is the linear deformation, φ_2 is the volume fraction of the polymer, χ is the Flory interaction parameter, and $\bar{V}_1$ is the molar volume of the swelling solvent. The slope of the line is well within the experimental error of -5/3! The calculation for the equilibrium swelling ratio was also successful. Goodyear was thrilled!

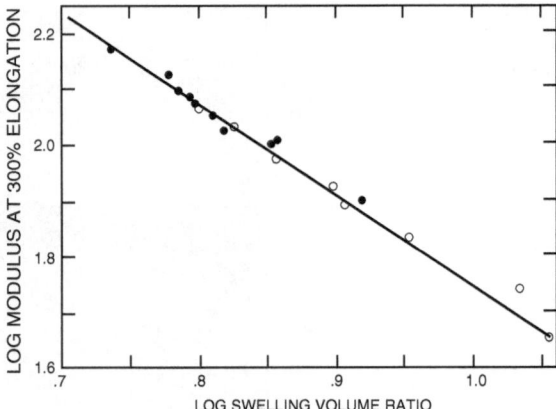

FIG. 8. Relationship between "modulus" and swelling volume ratio ($1/v_2$) for various
pure gum Butyl vulcanizates. $\circ$ = low unsaturation series; $\bullet$ = high unsaturation series.

Fig. 2.3 Elongational modulus as a function of swelling [21] (Chemical Reviews, by
permission)

Flory was also tapped to produce a critical review article on "Condensation
Polymerization." It became the basis for his Baker Lectures at Cornell and for
several chapters in his book, "Principles of Polymer Chemistry."

Ever eager to make actual progress in understanding, Flory considered the
serious discrepancy between the Flory-Huggins theory and actual data when the
concentration of polymer was below a critical concentration, now known as c*.
Under these conditions, the polymer concentration inside the **pervaded volume** of
individual chains was c*, while it was 0 elsewhere. Thus, the initial slope of a plot
of osmotic pressure against concentration was far lower than predicted by the
"mean-field" theory. Even the curvature was incorrectly predicted. It was time for
a better theoretical approach. The initial paper in this series was written at
Goodyear, and clearly makes the case for the phenomenological understanding; it
even achieves a modest agreement with experiment by using a perturbation theory
of interaction between polymer coils [22].

Another phenomenon that was of interest to Goodyear was the strain-induced
crystallization of natural rubber. It was another chance to apply both the theoretical
expertise developed by Flory over the last decade and the deep intuition gained
from experimental work in the industrial environment. (Flory once demonstrated
for me how to determine the average molecular weight of bulk polyisobutylene
from the tackiness of the liquid sensed with his own hand.) Melting points of
liquids are functions of pressure, and the quantity (dT_m/dP) depends on the sign of
the volume change on melting through the Clapeyron equation. The melting
temperature can also be manipulated by changing the entropy of the liquid phase
through orientation. Since the chemical potential of the liquid phase is increased
relative to the crystalline phase, the melting temperature is raised. Flory derived a

Fig. 2.4 Degree of
crystallization as a function
of elongation at different
temperatures [23] (J. Chem.
Phys., by permission)

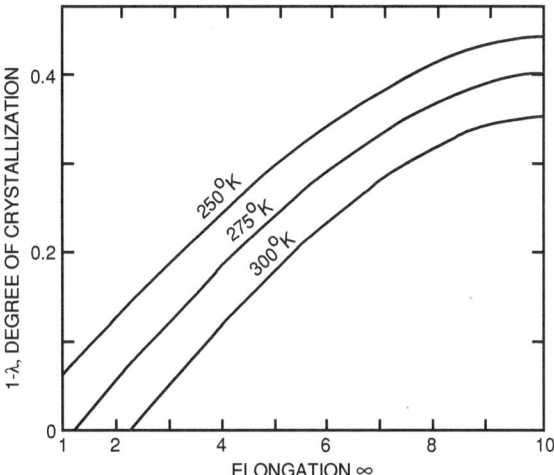

FIG. 1. Degree of crystallization $(1-\lambda)$ at equilibrium
. elongation at the three temperatures indicated. Curves
!culated from Eq. (18) taking $n=50$, $h_f=600R$ and
.$^{.}=250°$K.

statistical mechanical theory for the entropy of the liquid phase under uniaxial
stretching. He then constructed a phase diagram for the equilibrium degree of
crystallinity as a function of elongation. The result is shown in Fig. 2.4. At the
lowest temperature, the crosslinked rubber crystallizes to some degree in the
relaxed state. Unlike a macroscopic phase of a pure liquid, the crosslinked rubber
does not completely crystallize in the relaxed state. The crosslinks are excluded
from the crystal lattice. As the sample is oriented by elongation, it can crystallize
further since the chemical potential of the liquid is increased. At the highest
temperature, the sample will only crystallize when sufficient elongation is produced. Clear theory; clear presentation; semi-quantitative agreement [23].

Thomas G Fox (1921-1977)

Work on the problem of crystallization also introduced Flory to one of his best
colleagues and friends, Thomas G Fox (1921-1977). Fox received his Ph.D. in
physical chemistry from Columbia University in 1943. After military service he
joined the Goodyear Tire and Rubber Company in 1946. His first joint paper with
Flory appeared in 1948. It was the first of many joint papers on the temperature
and molecular weight dependence of viscosity.

Cornell University

The Cornell University Chemistry Department had a marvelous tradition that improved the intellectual environment of the place and served as a recruiting tool for the best people. In 1948, Peter Debye (1884-1966) invited Paul Flory to be the George Fisher Baker Non-resident Lecturer. In addition to giving a series of lectures, he brought along Tom Fox to continue their highly productive collaboration. Debye was one of the most famous scientists in the world and received the Nobel Prize in Chemistry in 1936. From 1934-1939 he was the Director of the Kaiser Wilhelm Institute in Berlin. But, in 1939 he accepted an offer to be the Baker Lecturer at Cornell, and never left. He became the Chair of the Chemistry Department at Cornell and dramatically improved the quality of research carried out there. With regard to polymer science, Debye developed the theory and technique of light scattering as a method for measuring the molecular weight and molecular size of polymers in solution. Flory also followed the tradition of remaining at Cornell as a Professor, his first real academic appointment.

William Krigbaum (1922-1991)

The academic position at Cornell allowed Flory to attract outstanding collaborators and Flory hired William Krigbaum (1922-1991) as a Postdoctoral Fellow. Krigbaum went on to have a brilliant career at Duke University. Flory's interest in the properties of dilute polymer solutions continued at Cornell and a major paper appeared in 1949 on the configuration of real polymer chains [24]. His work on this problem was greatly aided by Debye, who had used the Gaussian chain model of Kuhn to calculate the single chain scattering function of light scattering, S(q). This problem, known as the "excluded volume" problem, can be approached from several perspectives. The first concept is to consider the change in the dimensions of the chain due to the obvious fact that two segments cannot occupy the same space. Flory distinguished two cases: 1) local interactions that do not depend on molecular weight, and 2) long range interactions of segments separated by many units along the chain contour. If the true molecular chain is idealized as a sequence of Z statistical subunits comprised of many mers, two subunits can share the same center-of-mass location. The local interactions will affect the chain dimensions, but it will be reflected in a change in the "statistical segment length" in the Kuhn model. The long range interactions will depend on the molecular weight, since the average density of statistical segments is a function of the number of segments in the chain as a whole. For N statistical segments distributed in space around the center–of-mass in a Gaussian function, the average density of segments for an "unperturbed chain" will scale as $N/R_G^3 \sim N^{-1/2}$. When it is realized that the actual average radius of gyration is increased since chains with small radii of gyration will have a higher interaction probability, the expansion factor, α, can be expressed as:

$$\alpha^5 - \alpha^3 = C Z^{1/2}$$

where C is a constant and Z is the number of subunits in the chain. In the limit of large expansion, the radius of gyration increases as $Z^{0.6}$.

Another concept is to treat each individual chain as a thermodynamic system subject to elastic energies of expansion and thermodynamic free energies of dilution by solvent. With both the theory of elasticity and the concept of the interaction of two subunits as a mixing phenomenon, the highly repulsive limit derived by the first method was extended to include solvents where the interaction energy for chain subunits was more favorable than the interaction with the solvent.

$$\alpha^5 - \alpha^3 = C(1 - 2\chi)Z^{1/2}$$

where the interaction parameter, χ, is from the solution theory. When the polymer molecular segments and the solvent are very similar, the interaction parameter is small and the previous result is obtained, but when the solvent becomes poor, the interaction parameter can become large enough that the chain has its "unperturbed" average radius of gyration. It is not that the chain subunits do not interact; it is equivalent to the Boyle point in a real gas where the repulsive and attractive forces balance. The "excluded volume" between statistical subunits vanishes at a particular temperature for the polymer-solvent pair; this is now called the Flory temperature or Θ.

The time for a viable interpretation of intrinsic viscosities of polymer chains in solution was at hand. Werner Kuhn published many of the key insights for this problem. Einstein calculated from classical hydrodynamics that the intrinsic viscosity

$$[\eta] = \lim_{c \to 0}((\eta - \eta_0)/\eta_0 c) = 2.5(4\pi/3)R_e^3/M$$

depends on the hydrodynamic volume of a sphere of radius R_e. Kuhn proposed that for a high molecular weight chain, the solvent inside the coil would be entrained by the local frictional interactions, and hence the equivalent sphere would be proportional to the cube of the radius of gyration for the chain. This is now called the non-free draining chain. Flory then used his new quantitative relationship for the dependence of the radius on solvent quality and chain length to predict that the intrinsic viscosity would scale as $M^{1/2}$ in a Flory solvent and up to $M^{0.8}$ in a "good" solvent [25]. This agrees with the experimentally determined intrinsic viscosities for many polymer-solvent pairs! While a few subtleties remained, the basic conceptual picture has remained a pillar of polymer science.

A formally exact theory for the osmotic pressure of a dilute polymer solution can be constructed along the lines developed for dilute gases [26]. The key quantity needed for this approach is the potential of mean force between two full polymer coils in solution, $U(R_{12})$. The osmotic pressure can be expressed in dilute solution in the form:

$$\pi/c = (RT/M_n)(1 + A_2 M_n c + \cdots)$$

where

$$A_2 = \left(\frac{N_A}{2M^2}\right) \int_0^\infty \left(1 - \exp\left(-\frac{U(R_{12})}{k_b T}\right)\right) 4\pi R_{12}^2 dR_{12}$$

The potential of mean force for the chains as a whole is proportional to the excluded volume for a pair of subunits, V_e. This potential is given by:

$$U(R_{12}) = 2k_b T Z^2 V_e \left(\frac{3}{4\pi\langle R_G^2\rangle}\right)^{3/2} \exp\left(-\frac{3R_{12}^2}{4\langle R_G^2\rangle}\right)$$

While the Flory-Krigbaum potential is not exact, it is in closed form and enabled them to evaluate the integral and obtain:

$$A_2 = \left(\frac{N_A Z^2 V_e}{M^2}\right)$$

This result does not quite agree with experimental data as a function of molecular weight, since A_2 decreases weakly at higher values of Z. The Flory-Krigbaum potential is not quite correct, and a mean field approach is not quite right. But, the major improvement in the understanding of the second osmotic virial coefficient obtained by them remains a monumental achievement.

T. G Fox continued his studies of the temperature and molecular weight dependence of the shear viscosity of polyisobutylene. A monumental paper appeared in 1951 in the *Journal of Physical and Colloid Chemistry* [27]. Fox is now listed as present at Cornell. A thorough examination over a very wide range in molecular weight revealed that at high enough molecular weight the isothermal shear viscosity at 217 C could be described by:

$$\log \eta = 3.40 \log M - 13.56$$

A similar expression was found for polystyrene and the graph is shown in Fig. 2.5. It would be many years before the 3.4 power law observed at high molecular weight was explained, but the phenomenological standard had been established.

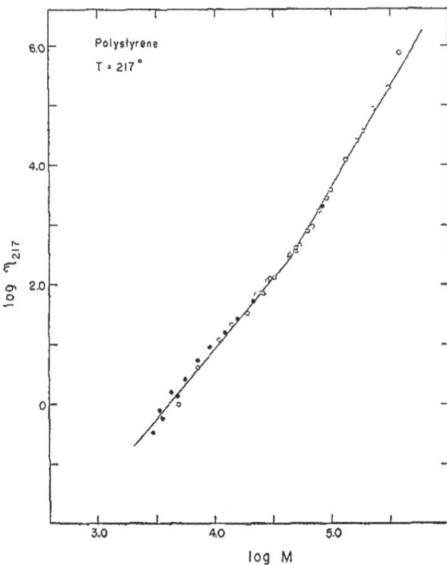

FIG. 10. Log η_{217} vs. log M for polystyrene fractions (6, 8). The straight lines correspond to equations 12 and 13.

Fig. 2.5 Log-log plot of viscosity against molecular weight [27] (Journal of Physical and Colloid Chemistry, by permission)

Another phenomenon associated with T. G Fox was the glass transition. A classic paper with Flory appeared in the *Journal of Applied Physics* in 1950 on the behavior of fractions of polystyrene as a function of temperature near the glass transition [28]. At this point, Fox and Flory referred to the observed phenomenon as a "second-order transition," a designation they would live to regret. Specific volumes were measured as a function of temperature on each fraction and the temperature where the thermal expansion coefficient showed a break in slope was taken as the glass transition temperature, T_g. An empirical correlation was constructed for this data.

$$T_g(K) = 373 - 1.0 \times 10^5/M$$

Equations of this type are now known as Flory-Fox equations. The quality of the data collected by Fox was so high that, like Faraday of old, it is still considered the best.

Cornell was a scientifically exciting place in the 1940s. Debye was leading a good corps of researchers in many directions. Since the random coil model was now an accepted paradigm, Debye explored the quantitative aspects of polymer chain configurations. He related the mean-square end-to-end distance and the mean-square radius of gyration, $\langle R_G^2 \rangle$. He even used this model to calculate exactly the single chain scattering function, now known as the Debye function:

$$S(q) = (2/u^2)[u - 1 + \exp(-u)]$$

where $u = q^2 \langle R_G^2 \rangle$. The scattering vector, $\vec{q}$, has a magnitude $q = (4\pi n/\lambda) \sin(\theta/2)$. The best values of the mean-square radius of gyration are obtained by measuring the entire function, not just the low angle limit.

One of the other Professors at Cornell during the 1940s was John Gamble Kirkwood (1907-1959). He developed an interest in the viscoelastic properties of fluids and applied his insights to polymer chains in solution. He applied the Oseen formulation of hydrodynamics rigorously to a chain composed of Z centers of frictional resistance. Kuhn's insight that the high molecular weight coil would entrain the pervaded solvent was given a quantitative basis. All the current work in single chain hydrodynamics is based on this pioneering work of Kirkwood [29].

Fox and Flory then applied the work of Kirkwood and Riseman to the problem of intrinsic viscosities in polymer solutions. They were able to use the single chain expansion theory of Flory and Krigbaum to derive a simple expression for the intrinsic viscosity in the high molecular weight limit:

$$[\eta] = \Phi \frac{\langle R^2 \rangle^{3/2}}{M}$$

where $\Phi = 3.62 \times 10^{21}$ is a universal constant, independent of polymer and temperature.

All the characteristics are then contained in the behavior of the mean-square end-to-end distance or the mean-square radius of gyration. In the good solvent limit, the intrinsic viscosity is predicted to scale as the 0.8 power of the molecular weight due to the strong expansion of the coil. For a globular coil, such as a protein, the intrinsic viscosity is independent of molecular weight. For a rodlike polymer, such as Staudinger proposed, the intrinsic viscosity is predicted to scale as the square of the molecular weight [30]. So much for the Staudinger viscosity law! The team of Kuhn, Debye, Bueche, Kirkwood, Riseman, Fox, Flory and Krigbaum established the actual paradigm for intrinsic viscosity. An immediate reward for Thomas G Fox was appointment as Director of Research at the Rohm & Haas Company in 1950 (Fig. 2.6).

Another of the major figures in polymer science to collaborate with Flory at Cornell was Arthur M. Bueche (1920-1981). He received his Ph.D. in Physical Chemistry with Debye on the conformational and frictional properties of polymers in solution. He became an expert on light scattering from polymer solutions.

Fig. 2.6 Thomas G Fox
(Carnegie Mellon University,
by permission)

DR. THOMAS G. FOX

A classic paper with Fox, Flory and Bueche on osmotic and light scattering data appeared in *JACS* in 1951 [31]. Bueche went on to become a star at the General Electric Company, a member of the National Academy of Sciences and was nominated as the Presidential Science Advisor.

Leo Mandelkern (1922-2006)

One of Flory's closest personal friends, Leo Mandelkern (1922-2006), joined him at Cornell in 1949 as a Research Associate after completing his Ph.D. at Cornell with Franklin Long. Mandelkern benefitted from interactions with Debye, Kirkwood, Scheraga and Flory. He became one of polymer science's leading experts on crystalline polymers, an area he pioneered with Flory. He went on to spend a decade at the National Bureau of Standards from 1952-1962. One area of special interest at Cornell was the use of frictional methods like diffusion, sedimentation and intrinsic viscosity to characterize polymer chains in solution [32] (Fig. 2.7).

With the torrent of papers flooding from the Flory laboratory, the time had come to prepare his Baker Lectures for publication. Much new material had appeared since 1948, and considerable advances in theory had been made. The unifying concept for

Fig. 2.7 Paul Flory, William Krigbaum and Leo Mandelkern at Cornell (James Mark, by permission)

the book is stated clearly in the Preface: "The author has been guided in his choice of material by a primary concern with principles." Hence the title: "Principles of Polymer Chemistry" [33]. While the scope of the book is vast, especially for 1953, Flory was careful to note that "it would scarcely be possible in a single volume to do justice to all the excellent researches in various branches of the subject." The fact that in 2014 it is still the central reference monograph in polymer science testifies to its importance. All of the issues raised in the preceding paragraphs are treated in detail in the book, and a uniform notation and point of view is adopted. This consistency and uniformity is one of the great strengths of the work.

References

1. Carothers WH (1940) Collected papers of Wallace Hume Carothers on high polymeric substances. Mark H, Whitby GS (eds), High Polymers Volume I. Interscience Publishers, New York
2. Carothers WH (1929) An introduction to the general theory of condensation polymers. J Am Chem Soc 51:2548–2559
3. Carothers WH (1931) Polymerization. Chem Reviews 8:353–426
4. Carothers WH, Williams I, Collins AM, Kirby JE (1931) A new synthetic rubber: chloroprene and its polymers. J Am Chem Soc 53:4203–4225
5. Carothers WH (1936) Polymers and polyfunctionality. Trans Faraday Soc 32:39–49
6. Flory PJ (1936) Molecular size distribution in linear condensation polymers. J Am Chem Soc 58:1877–1886

7. Kuhn W (1936) Kolloid Z 76:258–271
8. Flory PJ (1937) The mechanism of vinyl polymerizations. J Am Chem Soc 59:241–255
9. Staudinger H (1936) Trans Faraday Soc 32:97–115
10. Dostal H, Mark H (1935) Z Physik Chem B29:299
11. Hounshell DA, Smith JK (1988) Science and corporate strategy: Du Pont R&D, 1902–1980. Cambridge University Press, Cambridge
12. Flory PJ (1940) Molecular size distribution in ethylene oxide polymers. J Am Chem Soc 62:1561–1565
13. Flory PJ (1941) Molecular size distribution in three-dimensional polymers: I, gelation. J Am Chem Soc 63:3083–3090
14. Flory PJ (1942) Constitution of three-dimensional polymers and the theory of gelation. J Phys Chem 46:132–140
15. Flory PJ (1941) Thermodynamics of high polymer solutions. J Chem Phys 9:660–661
16. Flory PJ (1942) Thermodynamics of high polymer solutions. J Chem Phys 10:51–61
17. Treloar LRG (1949) The physics of rubber elasticity. Oxford at the Clarendon Press, Oxford
18. Gee G (1947) Some thermodynamic properties of high polymers and their molecular interpretation. Quarterly Rev Chem Soc 1:265
19. Flory PJ, Rehner J (1943) Statistical mechanics of cross-linked polymer networks: I, rubberlike elasticity. J Chem Phys 11:512–520
20. Flory PJ, Rehner J (1943) Statistical mechanics of cross-linked polymer networks: II, swelling. J Chem Phys 11:521–526
21. Flory PJ (1944) Network structures and the elastic properties of vulcanized rubber. Chem Reviews 35:51–75
22. Flory PJ (1945) Thermodynamics of dilute solutions of high polymers. J Chem Phys 13:453–465
23. Flory PJ (1947) Thermodynamics of crystallization in high polymers: I, crystallization induced by stretching. J Chem Phys 15:397–408
24. Flory PJ (1949) The configuration of real polymer chains. J Chem Phys 17:303–310
25. Flory PJ, Fox TG (1951) Treatment of intrinsic viscosities. J Am Chem Soc 73:1904–1908
26. Flory PJ, Krigbaum WR (1950) Statistical mechanics of dilute polymer solutions: II. J Chem Phys 18:1086–1094
27. Fox TG, Flory PJ (1951) Further studies on the melt viscosity of polyisobutylene. J Phys Colloid Chem 55:221–234
28. Fox TG, Flory PJ (1950) Second-order transition temperatures and related properties of polystyrene: 1, influence of molecular weight. J Appl Phys 21:581–591
29. Kirkwood JG, Riseman J (1948) J Chem Phys 16:565–573
30. Fox TG, Flory PJ (1949) Intrinsic viscosity-molecular weight relationships for polyisobutylene. J Phys Colloid Chem 53:197–212
31. Fox TG, Flory PJ, Bueche AM (1951) Treatment of osmotic and light scattering data for dilute solutions. J Am Chem Soc 73:285–289
32. Mandelkern L, Krigbaum WR, Scheraga HA, Flory PJ (1952) Sedimentation behavior of flexible chain molecules: polyisobutylene. J Chem Phys 20:1392–1397
33. Flory PJ (1953) Principles of Polymer Chemistry. Cornell University Press, Ithaca

Chapter 3
Herman Mark and Friends

Herman Mark (1895-1992)

High Polymers: Volume II "Physical Chemistry of High Polymeric Systems"

After the triumphant Faraday Society Meeting of 1935, Herman Mark (1895-1992) went back to Vienna and started preparing a monograph summarizing the current paradigms in polymer science. He also prepared his exit from Europe. His initial landing place was at the Canadian International Paper Company in Hawkesbury, Ontario. With the help of DuPont, Mark moved to the Brooklyn Polytechnic Institute in 1940 as Adjunct Professor of Organic Chemistry [1]. In 1946 he founded the Polymer Research Institute at Brooklyn "Poly." This center of polymer research was especially productive and the entire character of future group work in polymers was influenced by this effort. He also founded the *Journal of Polymer Science* in 1946 in collaboration with his publisher, Eric Proskauer (1903-1991), at Interscience Publishers [2].

His monograph, "Physical Chemistry of High Polymeric Systems", was published as Volume II in the Interscience Series on High Polymers in 1940 [3]. Mark wanted to set polymer science fully in the context of physical chemistry. This paradigm has continued to the present. He was an expert on chemical structure and presented the state of the art in both qualitative and quantitative structural chemistry. He seemed to know all the main players in this art, and was accepted fully as one of them. A good subdiscipline needs to maintain both the contact and the respect of the primary scientific community (Fig. 3.1).

Since polymer molecules are characterized by enormous numbers of structural states, the methods of statistical mechanics are required. Mark enthusiastically adopted this perspective for molecules in solution. He joined in the adoration of the "Gaussian chain," a random coil where the distribution of end-to-end lengths was a Gaussian function. The almost instant successful application to rubber

Fig. 3.1 Herman Mark
(1943) [7] (Interscience, by
permission)

elasticity cemented the validity of the model, just like the observation of the microwave background radiation validated the Big Bang Theory! No self-respecting polymer scientist can ignore the field of statistical mechanics.

Mark was one of the European experts on X-ray crystallography. He displayed his virtuosity in his chapter on crystal structure and its explanation in terms of interatomic interactions. For chain molecules, the repeating structure of the polymer was the first key concept. The chain conformation in the crystalline state needed to be regular and repeating. The chains are then packed into a unit cell. The richness of the number of possible unit cells for polymers was explained. Some polymer crystals actually have fully extended chain molecules in the lattice. The unit cell for this type of material is very large. But for much higher molecular weight materials with less than perfect crystallinity and with chain folded crystals, the unit cell reverts to a small unit. Globular proteins also crystallize, but the chains are compact. Even so, they have enough geometric particularity to arrange themselves into unit cells involving, for example, 12 protein molecules. Modern X-ray crystallography of such systems has reached a very high level of precision, with all the atomic positions, except for hydrogen, being determined. Even in 1940, smaller proteins, like insulin, pepsin and hemoglobin were successfully studied.

Herman Mark was also one of the leading liquid physicists, along with Prins, Debye, Frenkel, Lennard-Jones and Zernike. He included a thorough discussion of the interpretation of X-ray scattering from liquids in terms of the radial distribution function. In keeping with the views of his community, he modeled the liquid as a collection of small crystals. While this paradigm is now anathema in the liquid physics community, it still lives in the backwaters of German and Russian polymer science. Mark also included a discussion of liquid crystals.

The longest chapter in the book is devoted to polymer solutions. Mark was well aware that the classical equations contained in thermodynamics books like Lewis and Randall were inadequate for the quantitative description of polymer solutions, but he did not know yet where to find more than an admission of error and a few qualitative observations. Nevertheless, his collaborations with K.H. Meyer helped to establish the experimental situation in some detail. Meyer was now well established at the University of Geneva as Professor of Organic Chemistry. A description of his Interscience monograph on "Natural and Synthetic High Polymers" (Vol. IV, 1942) will be found below [16].

There is a thorough discussion of the osmotic pressure in polymer solutions. Due reverence is expressed to van't Hoff, but polymer solutions deviate from the asymptotic law at very low concentrations. One concept current in the 1930s was to introduce an "excluded volume", just as in the van der Waals theory of real gases. A standard experimental analysis of the data included representing the osmotic pressure in terms of a virial expansion. The phenomenological second virial coefficient was found to be a strong function of temperature in some systems. On the theoretical side, it was mostly groping in the dark!

Herman Mark was also a highly visible member of the solution viscosity community. Staudinger had staked out a leadership position with his "Viscosity Law," but since it was both experimentally and theoretically wrong, Mark introduced the next generation of workers to this field. (Staudinger claimed that the intrinsic viscosity of a chain molecule increased linearly with its chain length.) Werner Kuhn was at the center of the theoretical activity. Two of his colleagues at Vienna make their appearance here: Frederick Eirich (1905-2005) and Robert Simha (1912-2008). Eirich carried out many of the seminal experiments on the viscosity of colloidal and polymer solutions. The Einstein viscosity law was experimentally verified by Eirich [4]. (As noted in Chap. 2, the Einstein Law proposes that the intrinsic viscosity is proportional to the hydrodynamic volume of the suspended particle.) He was educated at the University of Vienna and received his Ph.D. with Wolfgang Pauli in 1929. Eirich had a brilliant career in Pauli's laboratory and illustrated the congenial relationship between colloid science and polymer science in Vienna. He joined Herman Mark's laboratory in 1934 after Mark returned to Vienna. His work was well received and was honored in 1938 by the faculty. By this time Herman Mark had escaped Austria by hiking on foot through the Alps into Switzerland. One of the great Mark stories concerns his method of transporting his wealth. He converted his fortune into platinum and hid it in the handle of his tennis racket. Mark helped Eirich to obtain a position with Rideal at Cambridge, which he started in 1938 (Fig. 3.2). He was able to obtain passage for his wife, Maria, and their daughter, Ursula, soon thereafter. Even his mother was allowed to leave Austria. Unfortunately, when Hitler annexed Austria, all Austrians were Germanized, and when the War broke out, all "Germans" in England were interned! To make matters worse, Fritz Eirich was sent to an internment camp in Australia, and only returned to Cambridge in 1943. After the War, Eirich and his family were granted a visa to emigrate to the United States. His mother arrived separately, and was promptly jailed because she was from Austria and spoke German!!! Eirich rejoined Mark at Brooklyn Poly in 1947 [5].

Fig. 3.2 Fritz Eirich and Sir Eric Rideal (Ursula Eirich Moeller, by permission)

Simha received a Ph.D. in theoretical physics from the University of Vienna in 1935, working with Eugene Guth (1905-1990). After Mark left Vienna, Simha fled to America and worked with Elliot Montroll at Columbia before rejoining Mark at Brooklyn Poly in 1941. After the War he joined the National Bureau of Standards (Fig. 3.3).

Attempts were also made to extend the Einstein law to higher concentration, both experimentally and theoretically. Eirich succeeded on the experimental side and Simha on the theoretical side. The work was extended to suspensions of ellipsoids and the dramatic increase in viscosity with elongation at constant volume fraction was derived by Simha and by Kuhn. The actual dependence of the intrinsic viscosity for a rod on length (and hence M) was as the square, not the linear dependence asserted by Staudinger. Another key figure in polymer science appears in this discussion: Maurice L. Huggins (1897-1981). A more thorough

Fig. 3.3 Robert Simha
(1912-2008) (Case Western
Reserve University, by
permission)

discussion of his contributions occurs below. Reference is also made to one of the
leading figures in the study of the mechanical properties of polymers, including
solutions: Roelof Houwink (1897-1988). And one of the editors of this series
appears as a central figure in the study of the viscosity of polymer solutions:
E.O. Kraemer (1898-1943) of DuPont. The advances detailed above in the
discussion of the work of Flory were yet many years away, but the worldwide
community of interest in the viscosity of polymer solutions was hard at work!

Two other dynamic experimental methods were discussed by Mark: diffusion
and ultracentrifugation. These fields benefitted from the leadership of The Svedberg
(1884-1971, Nobel 1926). While he was not primarily involved with the industrial
polymer community, he was a strong supporter from the beginning and a good
friend at all times. A strong research community needs a large group of scientific
friends who are favorable to the dominant paradigm of the community and who
provide advice and occasional guidance at key moments. His 1940 monograph with
K.O. Pedersen, "The Ultracentrifuge," is still the key document for this field [6]. A
good discussion of the application of ultracentrifugation to polymers by Elmer
Kraemer is also contained in another edited volume "The Chemistry of Large
Molecules" [7] (Fig. 3.4).

This Interscience volume(II) finishes with a few remarks about the kinetics of
polymerization, but, since volume III focuses on this topic, the next book by Mark
and Raff will be considered immediately following.

Fig. 3.4 Elmer O. Kraemer
(1898-1943) [7]
(Interscience, by permission)

High Polymers: Volume III "High Polymeric Reactions"

The key monograph in existence on the topic of polymerization in 1937 was by Robert E. Burk (1901-1978) (Western Reserve University), Howard E. Thompson (1893-1975) (Harshaw Chemical Company), Archie J. Weith (1886-1980) (Bakelite Company) and Ira Williams (1894-1977) (DuPont) [8]. Burk and Kraemer joined Mark in recommending that a volume on "High Polymeric Reactions" immediately be published in the Interscience Series on High Polymers. Mark enlisted R. Raff (1890-1975) from the Howard Smith Paper Mill in Canada to help him produce what became Volume III (1941) [9]. Raff was from the Vienna polymer community and had already published many papers with Mark and other leading figures.

The paper given by Herman Mark at the 1935 Faraday Discussion was on polymer kinetics [10]. This work formed the basis for his introductory chapters. The purpose of the initial chapters was to provide a sound paradigm for discussing the mountain of experimental work on polymerization that was accumulating. It also had a quantitative focus, as befits a physical organic chemist.

The necessary requirement for a quantitative description of polymer kinetics was a set of truly quantitative experimental methods. These form the basis of Chapter B. The field of reaction kinetics had achieved a very high level by 1941. Mark summarized it in Chapter C. The key references included books by Hinshelwood [11], Polanyi [12], Moelwyn-Hughes [13], Semenov [14], and Melville [15]. Considerable work had been carried out by H. Dostal and Mark in Vienna. While much of this work was superseded by Flory, it was a good contribution in its time.

"The General Theory of the Mechanism of Polyreactions" was the subject of Chapter D. Groundbreaking work by Ostromysslensky (1880-1939) in the early 20th century was followed through to 1941. The contributions of G.S. Whitby (1887-1972) were extensive. Both Dostal and Raff had published with Mark in this area. Carl S. Marvel (1894-1988) was a dominant figure during the period 1935-1953. Another member of the central European polymer community that emerged in this era was G.V. Schulz (1905-1999).

In order to begin to understand chemical kinetics at a fundamental level, quantum chemistry was necessary. A good introduction to the use of quantum ideas in the description of reaction mechanisms was presented. Polymer science continued to attract theorists in this era and F.T. Wall (1912-2010) is noted.

Every known quantitative study of polymerization is surveyed in Part II of this volume. It seems that the 1930s was one of the great ages of the detailed study of polymerization kinetics.

The march through the polymer zoo started with ethylene. Heating pure ethylene produces a *gemische* of organic substances. An extensive history of these efforts is given, but a glimmer of polymer science appears in a note about some strange solid material created by I.C.I. during very high pressure reactions. The discussion of propene introduces significant work by C.S. Marvel. Interestingly, the very old work of Marcellin Berthelot (1827-1907) was still being quoted in 1939. He may have usually been wrong, but he was very busy.

The polymerization of the higher olefins to macromolecules was achieved by catalytic polymerization by S.V. Lebedev (1874-1934) and many coworkers. His legacy in polymer science was carried on by N.N. Semenov (1896-1986, Nobel 1956) and by S.S. Medvedev (1891-1970). The Russian tradition in polymer chemistry is early and important. One of the most notable of these chemists, V.N. Ipatieff (1867-1952), immigrated to the United States in 1931. He was elected to the US National Academy of Sciences in 1939. He established an important research laboratory at Northwestern University for the study of high pressure and catalyzed reactions (Fig. 3.5).

Another polymer scientist that appears frequently in discussions of polymerization is Howard W. Starkweather (1914-) of DuPont.

Karl Ziegler (1898-1973, Nobel 1963) revolutionized the polymerization of olefins by the use of alkali metals and alkali-alkyls. His works on polybutadiene and polyisoprene were great examples of paradigm defining work.

One of the most interesting scientists that appears in the discussion of polyisoprene is James B. Conant (1893-1978) of Harvard. Although this was a very minor part of Conant's overall research program, it helped to inspire Wallace Carothers to pursue polymer science!

Advanced techniques, such as Hg sensitized polymerization, were employed by Sir Harry W. Melville (1908-2000, FRS) in the study of polyacetylene. Melville was a stalwart in British science, and contributed many ideas to the paradigm for polymer kinetics. He carried out his work at the University of Aberdeen (1945-1948) and at the University of Birmingham (1948-1956). He was honored with the Bakerian Lecture for 1956 on "Addition Polymerization" (Fig. 3.6).

Fig. 3.5 V. N. Ipatieff

Fig. 3.6 Sir Harry Melville,
K.C.B., FRS (Royal Society,
by permission)

One of England's greatest gifts to the United States was Sir Hugh Stott Taylor (1890-1974, FRS). He joined Princeton University upon receiving his Ph.D. in 1914 and became Chair of the Chemistry Department from 1926-1951. He had studied with both Bodenstein and Arrhenius. He carried out many investigations of the kinetics of polymerization.

There is a very extensive discussion of the thermal polymerization of styrene. This system has baffled the best scientists for years. What is now known is that in the liquid, but not the gaseous state, styrene starts polymerizing as soon as it condenses in a distillation flask. At low temperatures, the molecular weight of the polymer is very high. At higher temperatures the molecular weight decreases since the rate of chain transfer increases. While Flory's theory of chain transfer is mentioned, most of the other speculations were also given space. There appears to be a critical nucleus of confusion present in the discussion.

Thermal polymerization of indene under very high pressure was reported by P.W. Bridgman, of high pressure fame, and J.B. Conant. Similar systems were also studied by Whitby and by Marvel. When the right time occurs, many ideas are in the air. Some of them give life to science. Another bright idea concerned poly(chloroprene). Once Carothers had pioneered this area, high level work was conducted by Melville on the kinetics and mechanism.

Since many polymers are derived from inorganic monomers, a thorough review of this literature is also presented. A classic paper by Melville, discussed in *A Prehistory of Polymer Science* [26], reported studies of the polymerization of phosphorus. Kurt Meyer presented a paper on the polymerization of sulfur in the same *Transactions of the Faraday Society* volume on Polymerization.

A short chapter on polycondensations is also included. In addition to the classic work of Carothers, significant work by Raff and Dostal is mentioned. In addition to the groundbreaking theoretical work discussed above, Raff mentions the experimental work of Flory on diethylene glycol-adipic acid polyesterification.

Kurt H. Meyer (1883-1952)

High Polymers: Volume IV "Natural and Synthetic High Polymers"

In spite of the difficulties occasioned by the war, the English language translation of the monograph by Kurt H. Meyer appeared in 1942 as Volume IV of the Interscience Series on High Polymers: "Natural and Synthetic High Polymers: A Textbook and Reference Book for Chemists and Biologists" [16]. While some books become dated very quickly, especially synoptic surveys of a whole field of work, the monograph by Meyer is still an essential reference work! Much of the detailed information on specific polymers occurs only here. Textbooks tend to focus on "principles" and later monographs feature the latest work on very specific topics. The mountain of very specific detail found in this volume has rarely been repeated.

Meyer was a consummate scientist. He knew all the molecular physics being developed in the 1930s. He was a master of very many physical techniques. And, he was Professor of Organic Chemistry! The book contains a brief introduction to some physical techniques (and some that have rarely been discussed since), but the meat of the book is a thorough survey of all known polymers! Meyer was a champion of the view that all polyfunctional entities could be formed into macromolecules. It is hardly a surprise that he starts his survey of "inorganic" polymers with sulfur. While gaseous sulfur and ordinary crystalline yellow sulfur are composed of S_8 rings, liquid sulfur is made up of a very large number of chains of different lengths as well as some rings. The subject of ring-chain equilibrium was eventually treated in detail by Flory, but Meyer was well aware of the issue. A very large number of inorganic polymer topologies are presented. Both platelet and network systems are discussed. The unifying paradigm for all this work is the importance of the covalent bond network. In addition, Meyer discusses various forms of network decoration, especially the existence of charged groups. He not only laid the groundwork for the discussion, he illustrated it with specific inorganic polymeric substances.

The organic polymers are introduced with the paraffinic hydrocarbons. The n-alkanes provide a homologous series of well-defined materials with known chain lengths. The crystal structure as a function of chain length was determined by C.W. Bunn [17]. Bunn was one of the leading chemical crystallographers in England and produced the classic monograph "Chemical Crystallography" [18]. The chain geometry and the unit cell dimensions are shown in Fig. 3.7.

While polyethylene adopts a planar zig-zag conformation in the crystalline state, polyisobutylene is a helix. This conformation allows the pendant methyl groups to have more freedom while maintaining a perfectly repeating structure. Most polymer chains are helical in the crystalline state.

Substituted vinyl polymers are industrially very important. Polyvinyl chloride and bromide are discussed. An important figure in American industrial polymer chemistry, Waldo Semon (1898-1999), is mentioned in connection with the successful industrial development of plasticized PVC (Fig. 3.8).

Another class of industrially important polymers, polyacrylates and methacrylates, is associated with the firm of Rohm and Haas. Otto Rohm (1876-1939) was both a highly trained organic polymer chemist and a great industrialist. In 1907 he teamed with Otto Haas (1872-1960) to found both a successful German and American Company (Fig. 3.9). I owe a personal debt of gratitude to Otto Haas's son, John. He endowed the Charles Price Fellowship in Polymer History at the Chemical Heritage Foundation.

As discussed in "A Prehistory of Polymer Science," [26] polystyrene has played a pivotal role in the intellectual development of polymer science. In the period from 1935-1953 it also became industrially important, and Dow

The structure of a paraffin molecule, according to the earlier investigations of Müller and Hengstenberg and the more recent work of Bunn,[25] is shown in Fig. 46. Figures 47 and 48 represent projections along the b- and c-axes, respectively.

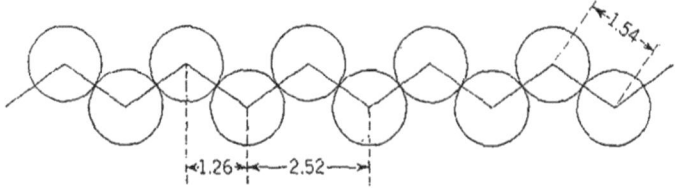

Fig. 46.—Carbon-atom chain of the paraffins.

From intensity measurements and a 3-dimensional Fourier analysis, Bunn obtained the section of a chain molecule along the chain-axis shown in Fig. 49, in which the contours unite points of the same electron density. The zig-zag structure is evident.

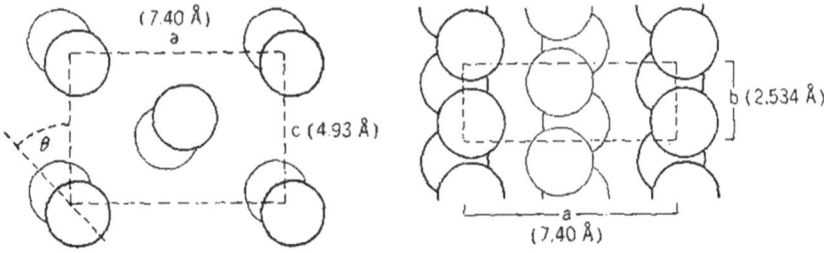

Fig. 3.7 Crystal structure of paraffins [16] (Interscience, by permission)

Fig. 3.8 Waldo Semon (Rubber Science Hall of Fame, by permission)

Fig. 3.9 Otto Rohm and Otto Haas (Rohm and Haas, by permission)

Chemical played a leading role. The classic monograph edited by Ray H. Boundy (1902-1992) and Raymond F. Boyer (1910-1993) on "Styrene: Its Polymers, Copolymers and Derivatives" was commissioned in this laboratory [19].

New developments in the understanding of polyisoprene also occurred in the timeframe of the current treatise, and Kurt Meyer was one of the leading workers in this area. Much better X-ray pictures were obtained from stretched sheets (Fig. 3.10). The good data allowed the intrinsic *cis* conformation of the double bond in natural rubber to be confirmed. It also yielded a very good model for the chain conformation as a whole in crystalline natural rubber and a deep insight into the unit cells observed. There are two common unit cells. The structure of crystalline gutta-percha was also studied successfully; the key step was knowing how to manipulate the sample into a highly oriented crystal. The double bond in gutta-percha is *trans*, but the crystalline helix can easily adopt two different forms, based on different rotation angles for the single bonds (Fig. 3.11).

(a)   **(b)**

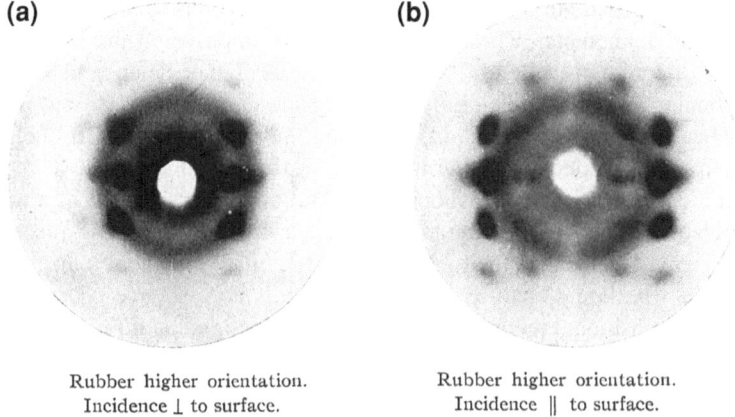

Rubber higher orientation.   Rubber higher orientation.
Incidence ⊥ to surface.   Incidence ∥ to surface.

Fig. 3.10 X-ray diffraction from stretched natural rubber films, both parallel and perpendicular to the film surface [16] (Interscience, by permission)

Fig. 3.11 The chain models for α and β gutta percha [16] (Interscience, by permission)

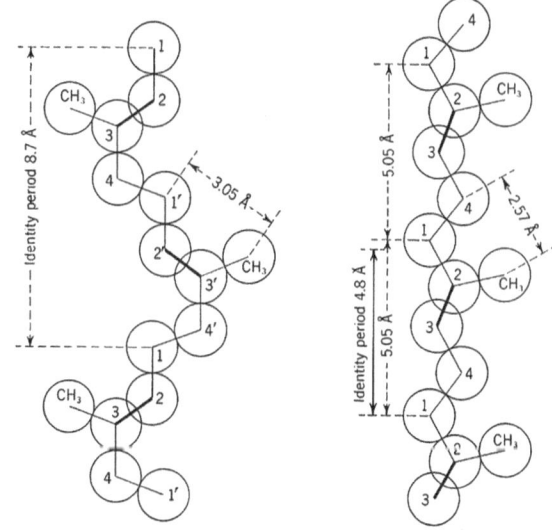

This work is fully modern and Bernard Wunderlich (1920-1988), the leading polymer crystallographer of the late 20th century, would be proud of it even now! Crystalline polymers can be discussed at many different levels: repeating chain conformation, unit cell structure and overall space group of full crystal. The polymorphic character of many polymer crystals was fully appreciated by Meyer.

Meyer also carried out thermo-elastic measurements on rubbery polymers using the Polanyi dynamometer. These were intended to improve on the seminal measurements of Joule, discussed in "A Prehistory of Polymer Science" [26]. Meyer held particular views with regard to the liquid state that have not fared well in the present, but they are still championed by some industrial scientists in the United States and some academics in Germany. He distinguished a unique "rubberlike state" in the behavior of natural rubber, and all rubberlike polymers, as a function of temperature: 1) at low temperature rubber crystallizes 2) above the melting point the sample then enters the rubberlike state 3) finally at higher temperatures it achieves the true liquid state. Even though there is no evidence of a thermodynamic phase transition, kinetic data is interpreted as a sign that a new phase exists. Even the best scientists sometimes abandon the safe haven of true thermodynamics for "the swamp of new phenomena." And bad science lives forever in the hearts of the true believers.

Meyer returned to sound thermodynamic analysis in the discussion of the dependence of the melting point on tension for lightly vulcanized gutta-percha. The classic curve is shown in Fig. 3.12.

Fig. 3.12 Tension dependence of the melting point of gutta-percha [16] (Interscience, by permission)

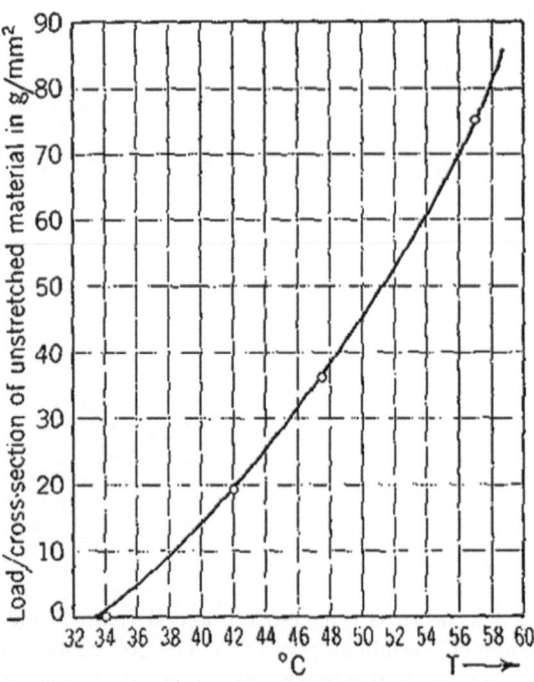

The melting point of gutta plotted
as a function of the tension.

The key thermodynamic relationship between elastic force, K, extension, melting temperature and entropy is:

$$K\left(\frac{\partial \Delta l}{\partial T_m}\right)_K = -\Delta S_m$$

Now this is Meyer at his best!

Meyer was also well aware that rubber was viscoelastic. Unvulcanized rubber continues to deform at constant small stress indefinitely above the melting region. Lightly vulcanized rubber deforms at a finite rate until it reaches elastic equilibrium. In addition to purely mechanical measurements, Meyer used birefringence as a probe of the degree of change in the sample. Rubber birefringence increases up to a limiting value over time at constant load. An even more interesting analysis takes into account that under high load, crystallization also contributes to the response and can even lead to a limiting deformation in unvulcanized rubber!

Meyer was also interested in synthetic rubber. At this time, no successful polymerization of isoprene led directly to the natural product. The macromolecules obtained were often branched and were not pure *cis*. Successful commercial rubbers were based on polymers of butadiene. The leader in this area was Karl Ziegler (1898-1973, Nobel 1963). The secret to making good polybutadiene was the right catalyst. Commercial rubbery polybutadiene (Buna) is observed to have the *trans* conformation at the double bond.

While there are commercial uses for pure Buna rubber, a more successful substance was created as a copolymer of butadiene and styrene (Buna-S). Another successful copolymer includes acrylonitrile (perbunan). Both products were widely available in Germany at the time of the writing of Meyer's monograph.

A specialized rubber was invented by Carothers at DuPont: poly(2-chlorobutadiene) or Neoprene. Because of the chlorine side groups, the rubber is resistant to hydrocarbons and can be used for fuel lines and other hoses that require solvent resistance. The double bond is observed to be *trans*.

Not all polymers have pure hydrocarbon backbones. The next material to be discussed was polyoxymethylene (POM). While Staudinger carried out many early studies of this material, Meyer was not averse to pointing out the large number of serious errors of interpretation associated with that work. POM is highly crystalline, but no discussion of the structure was included in Volume IV. Poly(ethylene oxide) is also readily synthesized. Crystal structure data was obtained by Hermann and Magda Staudinger. The interpretation is fanciful; perhaps they were rehearsing for the autobiography!

Another structural motif is the polyesters. They can be synthesized from either hydroxyacids(difunctional monomers) or as copolymers of diacids and diols. Poly(lactic acid) is one example of a difunctional monomer, and ethylene glycol/succinic acid is an example of the alternating copolymer form. Similar polymers can be formed as polyamides. While Nylon was big in the United States, it was only a footnote here.

Polymers involving sulfur in the backbone were studied by C.S. Marvel. A separate treatment of the work of Marvel is given below.

There is a section on Bakelite, and the work of N.J.L. Megson is noted. While many things were known about phenol-formaldehyde polymers in 1942, many of them are not mentioned here.

Another class of polymers that occupied 165 pages in the book and many years of effort by Meyer and his collaborators is the celluloses. The detailed X-ray crystallography that resolved the local and mesoscopic structure of cellulose and its micelles is presented in detail. His chief collaborators included Herman Mark, J.R. Katz, Michael Polanyi, L. Misch and G. von Susich. A detailed model of the unit cell of native cellulose is given in Fig. 3.13.

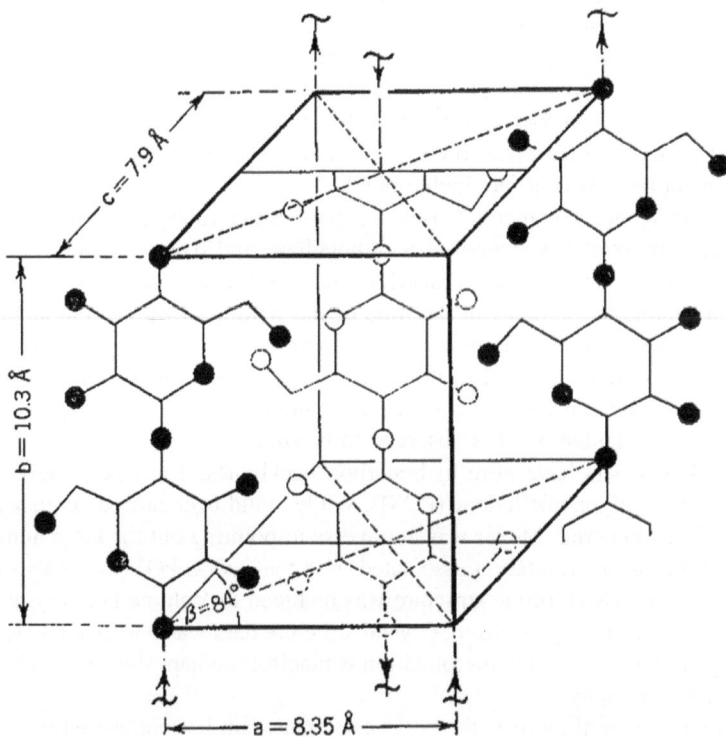

Diagrammatic representation of the unit cell of native
cellulose (Meyer and Misch).

Fig. 3.13 The unit cell of native cellulose [16] (Interscience, by permission)

Trichite structure of starch grain (A. Meyer).

Fig. 3.14 Crystalline morphology of starch [16] (Interscience, by permission)

The reactions of cellulose, mechanical properties of cellulose fibers, and the chemistry of cellulose degradation are discussed in detail.

This section is followed by another extensive discussion of starch, the other major commercial polysaccharide. It is amazing how much difference a simple stereochemical variation can make in the structure and properties of a macromolecule. Even more fascinating patterns are observed in native starch grains. The crystalline fibrils are arranged in a fractal-like pattern called trichites (Fig. 3.14).

Meyer was interested in all the natural polymers. The next group was the proteins. They are based on a set of amino acid monomers. The number of possible protein molecules is enormous. In addition to the seminal work of Meyer and Mark, W.T. Astbury (1898-1961) played a central role in the growing understanding of the structure of the different proteins in the period 1935-1953. The ultracentrifuge of Svedberg also was useful in the preparation of pure samples of specific proteins and The Svedberg was a world leader in protein science.

One protein that yielded to X-ray analysis was silk fibroin. Meyer and Mark were able to calculate the complete unit cell! The main chain conformation of collagen was also discovered. Because of the commercial importance of wool, the study of keratin fibers was heavily funded and was also successful.

Kurt Meyer was a leader in protein science and gave a 141 page review of the field, but major advances were soon to be made by others.

The book concludes with several synoptic chapters on polymer solutions, polymer films, and biological tissues.

The German edition of "Meyer and Mark" continued to be issued as one volume [20] (Fig. 3.15).

MAKROMOLEKULARE
CHEMIE

EIN LEHR- UND HANDBUCH FÜR CHEMIKER UND BIOLOGEN

VON

KURT H. MEYER UND H. MARK

ZWEITE AUFLAGE, VÖLLIG NEU BEARBEITET

VON

KURT H. MEYER

UNTER MITWIRKUNG VON A. J. A. VAN DER WYK

MIT 229 FIGUREN IM TEXT

1 9 5 0

AKADEMISCHE VERLAGSGESELLSCHAFT GEEST & PORTIG K.-G.
LEIPZIG

Fig. 3.15 Title page of the classic monograph by Meyer and Mark [20]

Turner Alfrey Jr. (1918-1981)

High Polymers: Volume VI "Mechanical Behavior of High Polymers"

The monograph on mechanical behavior was written by Turner Alfrey, Jr. (1918-1981) and published as Volume 6 of the Interscience Series on High Polymers in 1948 [21]. At the time he was an Assistant Professor of Polymer Chemistry at the Polytechnic Institute of Brooklyn. He joined Dow Chemical in 1950 and spent his career there. Alfrey received his Ph.D. in Polymer Chemistry from Brooklyn Poly in 1943. He returned to Poly in 1945 after a stint with Monsanto (Fig. 3.16).

Polymer science is a multidisciplinary community. While it is united by a love of macromolecules, each member also brings expertise from another community. The community of X-ray crystallographers contributed many of the early leaders in polymer science. Turner Alfrey was a member of the Society of Rheology and received its highest honor, the Bingham Medal, in 1954. In this monograph he combined his two loves in a monumental exposition of both rheology and polymer science.

Since most chemists and physicists have no background in rheology, it was necessary to give a thorough introduction to the principles of the subject. The classical limits of a perfectly elastic solid and a Newtonian liquid were extended to

Fig. 3.16 Turner Alfrey Jr. (1918-1981) (National Academy of Engineering, by permission)

include the actual case of viscoelasticity observed for all polymeric materials. The importance of treating both finite deformations and non-Newtonian flow was stressed. Deformation can alter the structure of a solid, just as flow can alter the local structure of the liquid, especially by orientation. The full free energy space of the material must be taken into account. Even phase transitions can be initiated by viscoelastic processes. All polymeric materials exhibit relaxation after the imposition of transient strain. The full time and frequency dependence of the viscoelastic response must be taken into account. For polymeric materials, the response function can exhibit change over 14 orders of magnitude in time. The phenomenological description of a viscoelastic material can be expressed in as complete a formulation as a theorist can tolerate. But, the microscopic interpretation of the macroscopic behavior was a challenge in 1948 and is a challenge today.

The elementary theory of the properties of solid atomic substances is expressed in terms of the forces between atoms and the motions of the atoms. Good agreement between measured mechanical properties of some simple solids and calculated compressibilities and shear moduli have been obtained. Even heat capacities have been calculated with accuracy. But, for solid polymers, the task is much more challenging. The microscopic explanation of liquid flow requires a detailed understanding of both the equilibrium structure of liquids and the modes of motion. Fortunately, Alfrey was influenced by liquid scientists like J.G. Kirkwood and Henry Eyring. A good discussion of the behavior of viscoelastic materials in the region of the glass transition is given. Another important influence on his thought was Robert Simha.

The second chapter considers the actual case of amorphous linear high polymers. It starts with 10 commandments for "plastoelastic behavior." The detailed understanding of the local chemical structure of polymers like polyethylene, polyisoprene or polystyrene leads to the conclusion that "a wide range of configurations are possible." This is due to the higher entropy associated with the many random configurations. Dynamic processes must be analyzed in terms of the local motions of chain segments and the overall diffusion of the whole molecule. Alfrey called these micro and macro-Brownian motion. Truly irreversible flow requires macro-Brownian motion. Local segmental motion can be associated with the "retarded elastic response" or configurational elasticity. He also distinguished a limiting elastic response at very short time that is determined primarily by the local liquid structure and the intersegmental forces. This paradigm is still in effect and can be summarized by the equation for creep compliance:

$$J(t) = J(0) + J_R(t) + t/\eta$$

The initial compliance $J(0)$ can be very small. The recoverable compliance $J_R(t)$ rises to a limiting value as the internal elastic processes reach their limiting strains. Finally, the material flows and the strain increases linearly in time with a slope determined by the true Newtonian shear viscosity.

Actual chain molecules display flexibility in the main chain conformation due to rotation about the single bonds. A good discussion for 1948 of this fact is given. Building on the foundational work of Guth and Mark [22] the statistical distribution function for the end-to-end length of the chain in the freely rotating chain limit is discussed. The Gaussian chain result of Kuhn is obtained.

In order to represent actual experimental data, idealized spring and dashpot models are constructed. Each spring is given a shear modulus and each dashpot a viscosity. In a simplified model with two springs and two dashpots, the instantaneous compliance is the reciprocal of $G(0)$, the time dependence of the retardational compliance is given by $(1 - \exp(-t/J_R\eta_R))$, and the ultimate flow is determined by the final dashpot viscosity. Actual retardation functions are more complicated! Alfrey was fully aware of the need for a distribution of retardation times. One of the most dramatic experimental evidences of the need for a distribution is the so-called "memory effect." Once the system leaves the equilibrium state, the properties of the system are determined by its actual time and temperature history. Since the different processes will have different compliances and different retardation times, it is possible to create states where competing deviations result in complex response functions, often overshooting the final equilibrium state. While this insight was developed in later years by many other workers, it is presented here in 1948!

The phenomenological language used here is the distribution of retardation times that determines the creep compliance $J(t)$. Various mathematical models are employed to model the observed behavior, but this is still early in the game for this approach. There is still limited good data, and it is well before the more sophisticated computer approaches of the future.

A systematic discussion is given of the actual experimental response of an amorphous polymer sample to particular thermomechanical histories. For a system subjected to a constant stress, S, the creep strain is given by:

$$\gamma(t) = \left[\frac{1}{G(0)} + \int_0^\infty J(\tau)(1 - \exp(-t/\tau))d\tau + t/\eta \right] S$$

where $J(\tau)$ is the distribution of retardation times. This is already the current paradigm. For a system subjected to a constant deformation, γ, the shear stress relaxation function is given by:

$$S(t) = \gamma \int_0^\infty G(\tau) \exp(-t/\tau)\, d\tau$$

where $G(\tau)$ is the distribution of shear relaxation times. Another popular mechanical history is to deform the sample at a constant rate, $d\gamma/dt$. The stress is then followed as a function of deformation.

$$S(\gamma) = (d\gamma/dt) \int_0^\infty \tau\, G(\tau)\, (1 - \exp(-\gamma/((d\gamma/dt)\tau)))\, d\tau$$

Finally, if the system is subjected to a time varying stress, the steady state response function for the shear strain is given by:

$$\gamma(t) = \left[\cos \omega t \int_0^\infty \frac{J(\tau)d\tau}{\omega^2\tau^2 + 1} + \sin \omega t \int_0^\infty \frac{\omega\tau J(\tau)d\tau}{\omega^2\tau^2 + 1} \right] S_0$$

I have stretched the reader's patience in including these equations so that the fact may be recognized that the modern era in polymer rheology has already arrived in the work of Alfrey! Everything afterward is the articulation of the paradigm.

Another class of polymeric materials is the three dimensional viscoelastic amorphous solids. They range from soft gels, which are two component systems, to hard Bakelite. Alfrey divided the materials into six arbitrary categories. What characterizes all of them is that they are solids; the shear viscosity is infinite and the static shear modulus is finite and measurable. They are also characterized by a distribution of relaxation times. The instantaneous shear modulus, $G(0)$, is characteristic of a typical liquid for most of the classes, but can achieve high values for Bakelite.

Since most rheologists would be unfamiliar with the microscopic theories of rubber elasticity, Alfrey presented a thorough discussion of the development of these equations. The key names include Kuhn, Guth and Mark, Wall, Treloar, Flory, and Guth and James. Separate treatments of the work of Wall and of Treloar will be found in another chapter. Alfrey is especially interested in the large strain limit, where the Gaussian theories are not appropriate. The inverse Langevin function appears here as a better representation of the chain when it is near full extension.

In spite of many arguments about theoretical artifices and the nature of the network in vulcanized rubber, actual force-extension data does not follow the classic equation of rubber elasticity over wide ranges in strain.

$$f(T, \alpha) = (Nk_bT/L_0)\left(\alpha - (1/\alpha)^2\right)$$

The actual network appears to have more elastically effective chains, N, than is predicted by the synthesis of the network when only data at small extensions is examined. At large extensions, the failure of the Gaussian approximation is clear and the forces also exceed the ideal prediction. An actual polymeric network composed of long chains crosslinked at random contains several types of network structures (defects). The apparent increase in the number of elastically effective chains has been attributed to the presence of chain entanglements, made effective by the real crosslinks, since they can no longer disentangle by Brownian motion but are trapped. In the intermediate range of extension, the force can become less than predicted; this has been attributed to intramolecular loops and to chain ends. This also implies that the trapped "entanglements" may become less effective as the chains align due to extension. In spite of the nuanced discussion in Alfrey, the polymer community chose to disintegrate into armed camps that featured ideo-logical fragility rather than viscoelastic response.

The distribution of retardation times is also modified by the presence of the network. Long wavelength modes are damped by the network and the distribution of chain lengths between crosslinks should lead to a softened cutoff at the long relaxation time end. This complexity suggests that actual equilibrium data should be taken for very long times, because full equilibrium may require extensive patience. Even if the network were "perfect" with identical numbers of monomers between crosslinks and no chain ends or loops, the physical location of the crosslinks would reflect the Gaussian statistics and the response would still be complex.

Alfrey includes a synopsis of the work of Flory on the formation of gels by polycondensation reactions [23]. He also recapitulates the discussion of Guth and James [24] on the explanation of the observed stress-extension curve.

When the number of crosslinks increases to the point where the chains between crosslinks are no longer in the Gaussian limit, the retardation spectrum changes dramatically. The glass temperature for the system is now approached in the experimental temperature window. Observed creep compliances are more like viscoelastic liquids. The breadth of the distribution is narrowed, compared to soft rubber, but is still several decades wide.

An extensive discussion of the stress relaxation function in vulcanized rubber completes this chapter. The work was performed by another emerging leader in polymer science: Arthur V. Tobolsky (1919-1972). He received his Ph.D. from Princeton University in 1944 and worked with Henry Eyring and Hugh S. Taylor. Taylor was proud to present the Bingham Medal of the Society of Rheology to Tobolsky in 1956. Tobolsky also collaborated with Herman Mark on the Second Edition of his monograph [3], published in 1950. He was so successful at Princeton that he was appointed there immediately. He found himself at Brooklyn Poly as Professor of Chemistry in 1950, but returned to Princeton where he spent the rest of his life. One of the first students to graduate under the direction of Arthur Tobolsky was Richard S. Stein (1925-) in 1948.

An extensive chapter on crystalline polymers set the paradigm for structure-property relations in this class of materials. A thorough understanding of the structure at all levels was stressed. The local molecular conformation, the unit cell, the microcrystalline structure, and the overall morphology, including both crys-talline and amorphous domains, were included.

In order to understand the final structure of a polycrystalline polymer sample it is necessary to consider its crystallization history. The rate of nucleation increases as the material is cooled below the thermodynamic melting point. At the melting point, the chemical potential of the liquid and the crystalline phase are equal, so that there is no driving force for a liquid to crystallize. The rate of crystal growth increases as the temperature decreases. The overall crystallization rate goes through a maximum well below the melting temperature. If the sample can be cooled rapidly enough, the crystallization regime can be avoided entirely and an amorphous glass can be created, but for most readily crystallized polymers the

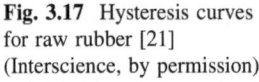
Fig. 3.17 Hysteresis curves
for raw rubber [21]
(Interscience, by permission)

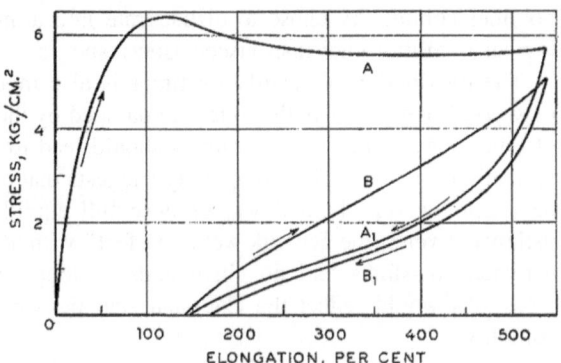

Stress–strain curves for raw rubber. A, B = strain curves.
A_1, B_1 = recovery curves.

thermal conductivity is low enough that this does not happen. Instead, the solid
sample contains many small crystallites. When such a sample is reheated to a
temperature just below the melting point crystallization can be almost explosive.
The collection of crystals of different size melts over a range in temperature, since
surface effects do influence melting for small crystals. Slow heating can result in
substantial change in morphology due to melting and recrystallization.

For polymeric samples, the imposition of external stress or strain can also
influence crystallization. The pressure dependence of the melting temperature is a
well-known effect predicted by the Clapeyron equation. For a rubbery liquid, the
imposition of extension leads to chain orientation that strongly influences the
melting temperature. A sample of crosslinked natural rubber crystallizes on
stretching at room temperature because the melting temperature is raised by the
orientation of the chains. Pure natural rubber crystallizes under no stress below
room temperature. Imposition of extension on such a material leads to further
crystallization and a major change in morphology. The importance of considering
all the variables that affect the free energy of the sample was stressed. Hysteresis is
the norm rather than the exception for such materials; this is illustrated by
Fig. 3.17. The complicated response to thermo-mechanical processing provides
both challenges and opportunities in industrial practice. Crystallization during
processing by stretching amorphous films produces tough, stable sheets.

The viscoelastic properties of plasticized polymers and polymer solutions are
also discussed in detail. This introduces another major figure in polymer science:
John D. Ferry (1912-2002). He took Stanford by storm and received both his
degrees by the age of 22. He was a Junior Fellow at Harvard. He arrived at the
University of Wisconsin in 1946, ready to vigorously pursue his career in research.
He was promoted to Full Professor the very next year!!! He received the Bingham
Medal (a lifetime award) in 1953, before both Alfrey and Tobolsky. A full dis-
cussion appears in Chapter 5.

The presence of low molecular weight plasticizer shifts both the time scale of molecular motion and the ultimate glass-like shear modulus at initial time G(0). The importance of frequency dependent measurements to the understanding of the viscoelastic properties of polymers was stressed.

The viscoelastic properties of dilute polymer solutions offer an opportunity to bring some rigor to both the theory and the experiments. In a dilute polymer solution, the individual chain molecules adopt a statistical ensemble of conformations which is constantly changing. This intramolecular Brownian motion is incessant. During flow, the molecules are subjected to forces that change the distribution of chain conformations and lead to measurable birefringence in the solution. The subject of the intrinsic viscosity immediately raises the question of the validity of the "Staudinger Law." The discussion by Alfrey leaves no doubt about the outcome: the "Law" is empirically unjustified and theoretically offensive.

Two of the early theorists that demolished the Staudinger Law were Kuhn and M.L. Huggins. Huggins derived a theoretical form that depends on the nature of the molecules in solution. This equation is now called the Mark-Houwink equation, but like many other historical artifacts, it traces its origin to Huggins.

$$[\eta] = K M^a$$

where the exponent, a, is 0 for a compact object like a globular protein, ½ for a Gaussian coil, and 2 for a rigid rod. Huggins also considered the influence of concentration on the measured solution viscosity, since for high polymers it is not possible to work at concentrations low enough to ignore interparticle effects.

$$(\eta_{sp}/c) = [\eta] + k' [\eta]^2 c$$

The Huggins coefficient, k', is observed to lie in the range 0-1. On the experimental side, Staudinger's own data refuted his Law. Flory and Mark also collected extensive data of intrinsic viscosity as a function of molecular weight. The observed exponent for random coil polymers in good solvents was in the range 0.6-0.8. There was still more to be done on this problem, but Alfrey summarized it brilliantly.

Another larger research community with strong representation in polymer science is the materials testing group. It is represented institutionally by the ASTM which defines measurable properties and oversees specific tests used in industry and government. One of the groups of materials properties of interest in the industrial use of polymers are the "ultimate properties." All materials fail when subjected to stresses in excess of those that can be sustained by the equilibrium structure. Polymers are no exception. If a polymer fiber is stretched too far or subjected to too large a tensile stress, it will break into two or more pieces. Prior to actual rupture, the sample may deform in an inelastic manner that radically changes the morphology of the sample. Alfrey was well attuned to this world. While its natural home is within industry, there are some scientists like Alfrey who brought critical thinking and a sound basis in physics and chemistry to the world of

Fig. 3.18 Tensile strength as
a function of molecular
weight [21] (Interscience, by
permission)

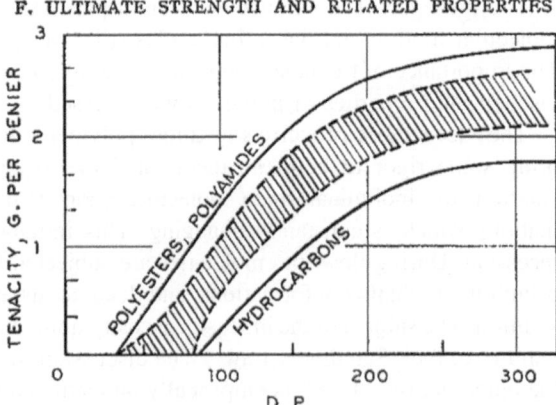

Tensile strength as a function of DP (after Mark).

ASTM. Typical arcana include tests like the "notched Izod impact test." Deciphering exactly what was being tested under this program was not trivial, and it required both rigorous physics and a strong stomach.

Polymer processing is a subtle balancing act between maximizing final properties within the constraint of actual processing time and conditions. High molecular weight polymers yield the best tensile strength, but the high viscosity at any temperature makes processing difficult. A lower molecular weight material can be processed, but may not stand up to actual use (Fig. 3.18).

From a strictly mechanical perspective, the question arises whether polydisperse materials yield better materials for industrial use, or whether monodisperse materials are superior. For many properties, it is the polydisperse sample that achieves the best performance. For other properties, a little low molecular weight impurity can seriously degrade properties like the ultimate modulus and the resistance to embrittlement over time. The goal of polymer engineering is to understand these issues well enough to blend the best brew.

One of the goals of polymer processing is often to produce highly aligned filaments. Another processing goal is biaxially oriented films. Alfrey discussed methods of evaluating both the orientation and the crystallinity of the samples before and after stretching. This type of detailed industrial analysis is carried out with all the rigor of "fundamental" science, but it often gets little acknowledgement from more mathematically inclined engineers.

Polymers are used over a wide range of temperature. The properties are often strong functions of temperature. Considerable industrial research into the range of observed behaviors is discussed here. Other polymers respond strongly to the relative humidity. Unless the material performs adequately under the range of expected conditions, it is useless for industrial practice.

Alfrey displayed both a high level of mathematical sophistication and a keen appreciation of the realities of industrial practice. In this regard, he was a model polymer scientist.

Fig. 3.19 Eric Proskauer
(1903-1991)

Eric S. Proskauer (1903-1991)

In addition to being a publisher with Interscience, Eric Proskauer was a beloved member of the polymer community. A touching tribute to his 70th birthday was published by John Wiley and Sons, who acquired Interscience [25] (Fig. 3.19).

Eric S. Proskauer was born in Frankfort am Main in 1903 and educated at Leipzig University. He worked with Carl Drucker and obtained his Ph.D. in 1931. He fell in love with editing and publishing and worked for Akademische Verlagsgesellschaft. In 1937 he immigrated to the United States, and founded Interscience Publishers with Marcell Dekker in 1940. They wasted no time in publishing the series on High Polymeric Substances!

The Tribute volume is filled with praise from his myriad friends. The polymer community owes Eric Proskauer a great debt of gratitude.

The story of Herman Mark and his polymer friends provides a strong basis for a coherent narrative. Brooklyn Poly was a fortress from which many polymer battles could be fought. With the death of Herman Mark, times have changed and the Polymer Research Institute is now an historical artifact.

References

1. Hounshell DA, Smith JK (1988) Science and corporate strategy. Cambridge University Press, Cambridge
2. Morris PJT (1986) Polymer pioneers. Beckman Center for the History of Chemistry, Philadelphia

3. Mark H (1940) Physical chemistry of high polymeric systems. High Polymers Volume II. Interscience Publishers, New York
4. Eirich F (1937) Kolloid Z. 81:7–18
5. Moeller UE Eirich (2011) Eirich family odyssey. Privately published, Santa Fe
6. Svedberg Th, Pedersen KO (1940) The ultracentrifuge. Oxford University Press, Oxford
7. Kraemer EO (1943) in The chemistry of large molecules, Burk RE, Grummit O (eds) Interscience Publishers, New York
8. Burk RE, Thompson HE, Weith AJ, Williams I (1937) Polymerization and its applications in the fields of rubber, synthetic resins and petroleum. ACS Monograph No. 75, Reinhold Publishing Company, New York
9. Mark H, Raff R (1941) High polymeric reactions: their theory and practice. High Polymers Volume III. Interscience Publishers, New York
10. Dostal H, Mark H (1936) The mechanism of polymerization. Trans Faraday Soc 32:54–69
11. Hinshelwood CN (1926) Chemical change in gaseous systems. Oxford at the Clarendon Press, Oxford
12. Polanyi M (1932) Atomic reactions. Williams and Norgate Ltd, London
13. Moelwyn-Hughes EA (1933) Kinetics of reactions in solution. Oxford at the Clarendon Press, Oxford
14. Semenoff N (1935) Chemical kinetics and chain reactions. Oxford at the Clarendon Press, Oxford
15. Farkas L, Melville HW (1939) Chemical kinetics. Cambridge University Press, Cambridge
16. Meyer KH (1942) Natural and synthetic high polymers. High Polymers Volume IV. Interscience Publishers, New York
17. Bunn CW (1939) Trans Faraday Soc 35:482–491
18. Bunn CW (1946) Chemical crystallography. Oxford at the Clarendon Press, Oxford
19. Boundy R, Boyer RF (1950) Styrene: Its polymers, copolymers and derivatives. Reinhold Publishing Company, New York
20. Meyer KH, Mark H (1950) Makromolekulare chemie. Akademische Verlagsgesellschaft Geest and Portig K.-G., Leipzig
21. Alfrey T (1948) Mechanical behavior of high polymers. High Polymers Volume VI. Interscience Publishers, New York
22. Guth E, Mark H (1934) Monatsh 65:93–121
23. Flory PJ (1944) Chem Revs 35:51–75
24. Guth E, James HM (1941) Ind Eng Chem 33:624–629
25. (1974) Eric S. Proskauer: A Tribute. John Wiley and Sons, New York
26. Patterson G (2012) A prehistory of polymer science. Springer, New York

Chapter 4
Speed Marvel and Friends

Carl Shipp Marvel (1894-1988) took the mantle left by Wallace Carothers. It is not surprising that he also came from the University of Illinois. The towering figure at Illinois in this period was Roger Adams (1889-1971). He was Department Head from 1916-1954. Adams was from a famous New England family and attended Harvard University.

After obtaining his Ph.D. in 1912, Adams spent two years in Europe on a Parker Traveling Scholarship in the laboratories of Emil Fisher and Otto Diels in Berlin and with Richard Willstatter at Berlin-Dahlem. He returned to Harvard for a few important years. He became friends with Elmer Keiser Bolton (1886-1968) (later of DuPont), Farrington Daniels (1889-1972) (later of Wisconsin), Frank C. Whitmore (1887-1947) (later of Penn State), James B. Sumner (1887-1955, Nobel 1946) (later of Cornell) and James Bryant Conant (1893-1978) (later President of Harvard).

In 1916 Roger Adams was invited by William A. Noyes (1857-1941) to come to the University of Illinois at Urbana-Champaign. Over the next 38 years he mentored more than 250 graduate students. He was the editor of "Organic Syntheses" and had a major impact on natural products chemistry. One of his best students was Wallace Carothers. Another chemist with whom he worked was Speed Marvel. Carl was known for being eager for breakfast and the nickname stuck.

Carl Marvel was born on a farm in Illinois. His family came to America in the 17th century! Farm life was important to Marvel and he developed a lifelong love of birds and flowers. He attended Illinois Wesleyan University and received both A.B. and M.S. degrees in Chemistry in 1915. He entered the Graduate College of the University of Illinois in 1915. He loved to synthesize organic compounds and when Roger Adams recruited some of the Illinois graduate students to join the war effort by synthesizing rare chemicals unavailable because of the war, Speed was quick to respond. He obtained his Ph.D. under William A. Noyes in 1920 and was immediately hired by Roger Adams as an instructor. He progressed through the ranks until 1961, when he needed to "retire."

Marvel absolutely loved organic chemistry and contributed substantially to "Organic Syntheses." He was a reliable expert on all that was happening in

G. Patterson, *Polymer Science from 1935–1953*, SpringerBriefs in History of Chemistry, 51
DOI: 10.1007/978-3-662-43536-6_4, © The Author(s) 2014

Fig. 4.1 Speed Marvel
(National Academy of
Sciences, by permission)

synthetic organic chemistry, and was tapped to consult at DuPont in 1928. One of
the consequences of his interactions with DuPont was his exposure to polymerization. He considered macromolecules to be the logical extension of his research
on simple organic molecules.

Marvel contributed to our understanding of scores of polymerizations. His first
large area of study involved the copolymerization of sulfur dioxide and olefins.
There are many structural issues to be considered in this system, as illustrated by
the following statement:

"Polypropylenesulfone was found to have a head-to-head, tail-to-tail orientation, while the polymer from vinyl chloride and sulfur dioxide was found to have a
head-to-tail array of two units of olefin to one of SO_2" [1].

He investigated the stereochemical structure of many vinyl homo and
copolymers. He endeavored to synthesize stereoregular polymers, but the time was
still early for this topic.

All his hard work paid rich dividends. He was elected to the National Academy
of Sciences in 1938 and served as the Chairman of the Chemistry Section from
1944-47. He received the Nichols Medal of the New York Section of the ACS in
1944. With this trajectory, it might have been expected that he would receive the
Nobel Prize, but not everyone who deserved the prize received it.

During World War II, Carl Marvel was of great service to the government. He
served in the Rubber Reserve Corporation and organized a team at the University
of Illinois to synthesize substitute rubbers. He received the President's Certificate

of Merit for Civilians in World War II. After the War he was part of the team that toured Germany to glean as many scientific and technical advances as they could.

He extensively studied sulfur compounds and their use in macromolecular chemistry. This love may have been developed during his younger days as a skunk trapper! The range and depth of his work was astounding, and by 1953 he was the leading synthetic polymer chemist in America (Fig. 4.1).

Calvin Everett Schildknecht (1910-2003)

Calvin Schildknecht was born in Gettysburg, Pennsylvania and died there as well, after teaching at Gettysburg College for 50 years. During the period of interest of this monograph, he was one of the well-known American synthetic polymer chemists. He was fortunate to have been influenced at Johns Hopkins by such stalwarts as Emil Ott and Maurice Huggins. This led, upon receiving his Ph.D. in physical organic chemistry, to a job at DuPont. There he interacted with consultants like Carl Marvel and Herman Mark. In 1943 he moved to the American subsidiary of BASF, now known as GAF (and fully independent of the former German parent company). One of his primary coworkers was Joseph Lambert, a former student of Herman Mark while still in Vienna.

While studying the polymerization of vinyl isobutyl ether, he stumbled upon a highly crystalline version, and created a sensation when he presented his work at the American Chemical Society Meeting at Atlantic City in 1947. He was greatly encouraged by the support of Maurice Huggins in this work. He had discovered highly stereotactic polymer! The full story of his work is summarized in his classic book from 1952, "Vinyl and Related Polymers" [2]. While he did not receive the Nobel Prize for this work, it was highly regarded in America [3].

Karl Zeigler (1898-1978)

Karl Ziegler was a towering figure in 20th century chemistry and received the Nobel Prize in 1963. He was educated at the University of Marburg and was appointed as Privatdocent at a very young age. He made his way up the German academic ladder and established a research institute at Halle. He worked in many areas of synthetic and physical organic chemistry, but he was best known for his groundbreaking work in organometallic chemistry.

The German scientific community has benefitted not only from a strong university system, but also from the existence of well-financed research institutes. Karl Ziegler was appointed as Director of the Max Planck Institute for Coal Research in 1943. In typical European tradition, he also retained his Professorship at Halle. With the advance of the Russian army in 1945, only a friendly rescue operation by an American Army truck saved his life. His days in Halle, and the

Fig. 4.2 Karl Ziegler
(1898-1978, Nobel 1963)

city itself, were doomed. Ziegler made fast friends with the Americans and helped in the rebuilding of Germany. Marshall Plan money helped to rebuild the Max Planck Institute in Mulheim. Ziegler was the most famous chemist in Germany after the war, and when the Allies allowed the German Chemical Society to reorganize in 1949, he was elected President.

Ziegler's scientific achievements were numerous and varied. He directed a large and loyal staff. But one of his best friends was Herman Mark, and when more normal relations resumed after the war, and it was again safe for Mark to travel in Europe, Ziegler asked Mark to "tell me a little about polymers." The rest is history! Using his newly synthesized aluminum alkyl catalysts, he was able to make linear polyethylene under low pressure and normal temperature conditions. Herman Mark was the much beloved "Geheimrat" of polymer science. He promptly told all his friends, including Sir Robert Robinson, about the new chemistry. Robinson promptly convinced his firm, Petrochemicals Ltd., to link up with Ziegler. The American firm that won the "Ziegler Prize" was Hercules Powder Company. Like DuPont, they wished to diversify after the war into more general chemicals, not just gun powder! (Fig. 4.2).

Giulio Natta (1903-1979)

Giulio Natta played a similar role to Ziegler, but in Italy. He was educated at the Politechnico di Milano and received his degree in chemical engineering in 1924. He achieved the Libero Docente degree in 1927 and was appointed as a professor at Pavia in 1933. He moved to Rome in 1935 and to Turin in 1936. In 1938 he returned to Milano as head of the chemical engineering department.

Like Ziegler, he was interested in the role of catalysts in chemical processes. He applied this knowledge to the synthesis of rubber and other polyolefins. Eventually he succeeded in making stereoregular polymers. In addition he was able to establish through X-ray crystallography the true structure of these materials.

Fig. 4.3 Giulio Natta (Nobel
Archives)

He was tightly associated with the Montecatini Company in Italy, and they both
funded his research and filed for the patents. He published more than 700 papers.
He received the Nobel Prize in 1963, along with Zeigler.

At one point, Ziegler and Natta were close collaborators, and shared
researchers. Eventually, the foreign "visitors" served more as industrial spies, and
the working relationship ceased. The whole story is charmingly told by Frank M.
McMillan (1914-2001), the Research Director of Shell, in his book *The Chain
Straighteners* [3]. The paradigm of engineered polymers with precisely known
structures was firmly established by Ziegler and Natta, and the number of such
polymers is now extremely large (Fig. 4.3).

Frederick T. Wall (1912-2010)

Fred Wall was born in Minnesota and educated there as well, receiving his Ph.D.
in 1937. As part of his Ph.D. research, he spent a year with Linus Pauling at the
California Institute of Technology. He was appointed to the University of Illinois
in 1937 and stayed until 1963, finishing as Dean of the Graduate College. He was
elected to the National Academy of Sciences in 1959.

The University of Illinois was a scientifically productive place for polymer
science. Carl Marvel exposed Wall to the many interesting problems that needed
theoretical treatment. He is especially known for his work on the theory of rubber

Fig. 4.4 Chemical
thermodynamics

elasticity. I was first aware of Wall's work through his thermodynamics textbook, one of the few that treated the full theory of rubber elasticity [4] (Fig. 4.4).

When Fred Wall was the Vice President for Research at Illinois, he was heavily involved in the creation of the legendary ILLIAC computer. He used it himself to calculate the statistical properties of chain molecules. He used the Monte Carlo method to assess the average conformations of the macromolecules.

Rudolph Signer (1903-1990)

One of the leading figures at the 1935 Faraday Discussion on Polymerization was Rudolph Signer from the University of Berne. He continued at Berne from 1935-1972 and made many important discoveries in polymer science. He was educated at the Swiss Federal Institute of Technology in Zurich and traveled with Staudinger to Freiburg in 1926. He played a major role in the research carried out in Staudinger's laboratory. His Ph.D. in 1928 was from Zurich and Emil Ott was on his committee.

In 1933 he was a Rockefeller Fellow at Uppsala with The Svedberg and at Manchester with Bragg! These experiences radically changed his perspectives and he did not return permanently to Freiburg. He was appointed Professor of Organic Chemistry at the University of Berne in 1935. Forever after, Signer applied the most rigorous physical methods to the study of polymers in solution. He is most famous for his studies of DNA and other nucleic acids using ultracentrifugation and flow birefringence. He was nominated for the Nobel Prize for this work [5].

References

1. Leonard NJ (1994) Carl Shipp Marvel 1894-1988. National Academy of Sciences, Washington, D.C.
2. Schildknecht C (1952) Vinyl and related polymers. John Wiley and Sons, New York
3. McMillan FM (1979) The chain straighteners. The Macmillan Press, London
4. Wall FT (1958) Chemical thermodynamics. W.H. Freeman and Company, San Francisco
5. Signer R (1986) Oral history. Chemical Heritage Foundation, Philadelphia

Chapter 5
Other Emerging Leaders

Many other scientists committed a major part of their work to the study and understanding of macromolecules. In this chapter, brief surveys of these emerging leaders are given.

Emil Ott (1902-1963)

Emil Ott was universally recognized as the "Cellulose Man." He was the one chosen to edit the Interscience book "Cellulose and Cellulose Derivatives", Volume V in the High Polymers Series [1]. And he was chosen to talk on Cellulose in the Western Reserve lecture series in 1942 [2] (Fig. 5.1).

Ott was born in Zurich and received his Ph.D. from the Swiss Institute of Technology in 1927. He immigrated to the United States in 1927 and worked at Johns Hopkins until 1933, when he joined the Hercules Powder Company in Wilmington, Delaware. He was the Director of Research at Hercules from 1939-1955.

Emil Ott was a firm believer in the team approach to research and he used this organization to solve an enormous number of industrial problems for Hercules. His approach to science was systematic and comprehensive. He adopted a complete structural description, like Haworth. He considered the detailed chemical analysis of his macromolecules. If the chemistry was wrong, the structure was wrong! He was an expert in X-ray analysis. The unit cell of crystalline cellulose can be constructed in two different ways, with two cellobiose units. Hydrogen bonding plays an important role in crystalline cellulose.

Ott was sensitive to the full range of industrial issues, as well as all the scientific ones. The nature of a cellulose sample was a strong function of its source and method of purification. It was also a strong function of the processing conditions and history. The crystal structure of "mercerized" cellulose is different than native fiber cellulose. Cellulose is often used as fibers. The morphology of cellulose fibers was one of Ott's specialties.

Cellulose is also the raw material for many modified polymers. Ott was an expert on cellulose reactions. Because of the multiple sites on cellobiose for

G. Patterson, *Polymer Science from 1935–1953*, SpringerBriefs in History of Chemistry, DOI: 10.1007/978-3-662-43536-6_5, © The Author(s) 2014

substitution, a partially derivatized sample has an enormous number of chemically different products. This heterogeneity can be both an advantage and a disadvantage, depending on the application. The commercial cellulose derivative of most interest to Hercules was "gun cotton," the nitrocellulose macromolecule. Ott led a team to optimize the production of gun cotton.

Ott was knowledgeable about both the fundamental chemical and physical issues and about the practical issues involved in industrial processing. Considerable effort was expended to study the optimal degree of substitution for particular applications. The understanding of the uses of plasticizers in processing was impressive. In substituted cellulose, they serve both as solvents and as softeners.

Emil Ott was justly celebrated for being the full industrialist: he knew the basic science, he knew his specific material, and he knew his market. He was very personable as well and was welcomed throughout the polymer world.

Maurice Loyal Huggins (1897-1981)

Maurice Huggins received training in the laboratory of G.N. Lewis at the University of California, Berkeley. He received his B.S., with a written thesis, in 1919, and his Ph.D. with Charles Porter in 1922. He was in the chemistry department at Johns Hopkins until 1936. His undergraduate interest in hydrogen "bridges" was fully developed and in 1936 he published a seminal article in the first volume of the Journal of Organic Chemistry on the topic [3].

He took a job with Kodak Research Laboratories in 1936. He developed an intense interest in the effect of hydrogen bridges on the structure of crystalline proteins and published a review article on this topic in 1943. He also produced a model of the protein alpha helix in 1943. While his contributions to the understanding of the conformations of proteins were early and important, he never received extensive recognition.

Huggins is best known in the polymer community for his work on the thermodynamics and rheology of polymer solutions. He was one of the developers of the Flory-Huggins theory of polymer solution thermodynamics. While Flory instinctively understood that the mean field approximation inherent in this approach limits it to concentrated solutions, Huggins always thought that it was "exact" in the dilute solution. He clearly missed the issue of the excluded volume! Huggins extended the work of Guth and Simha on solution viscosity. The correction for finite concentration is traditionally expressed in terms of a quantity known as the Huggins coefficient.

Raymond Matthew Fuoss (1905-1987)

Raymond Fuoss was a 20th century polymath who spoke 19 languages fluently. He entered Harvard at 17 and had published extensively by the time he graduated in 1925. He was a Sheldon Fellow at the University of Munich and worked with Wieland, Fajans and Lange. Eventually he entered Brown University to work with Charles A. Kraus, the famous electrochemist and ACS President. He graduated in 2 years after writing his actual thesis with Lars Onsager on irreversible processes in non-aqueous solvents. He remained at Brown until 1935 when the General Electric Company made him an offer he could not refuse.

Fuoss was one of the most educated chemists in America, and he applied both his knowledge and his brilliant theoretical mind to the problem of the electrical properties of polymers. He established the paradigms that still undergird research in this area. He was chosen to speak at the Western Reserve Lectures in 1942 on this topic [2] (Fig. 5.2).

Fig. 5.2 Raymond Fuoss [2]
(Interscience, by permission)

Once the war was over in 1945, Fuoss accepted the Sterling Chair in Chemistry at Yale. The team of Onsager, Kirkwood and Fuoss made Yale the premier theoretical chemistry department in the world at this time. The focus of his work at Yale was on electrolyte solutions, polyelectrolytes and the statistical mechanics of condensed phase systems. He was elected to the National Academy of Sciences in 1951. While his interests continued to evolve after this period in directions different than the community of polymer scientists, his contribution to the period of the consolidation of the basic paradigms was seminal. The unique phenomena associated with linear polyelectrolytes stimulated much theoretical activity, both in the time of Fuoss and afterward. It remains a challenging, but rewarding, topic today.

Leslie Ronald George Treloar (1906-1985)

L.R.G. Treloar was born in the United Kingdom and educated at the University of Reading (B.Sc. 1927, Ph.D. 1938). During the years between his undergraduate degree and his doctorate he worked at the General Electric Company Limited at Wembley. Following his doctorate he joined the British Rubber Producers Research Association (BRPRA). His assigned task was to understand the physics of rubber elasticity. The next ten years were devoted to an intensive study of rubber, culminating in the classic monograph *The Physics of Rubber Elasticity* [4]. He was joined in many of his studies by Geoffrey Gee (Fig. 5.3).

Treloar studied all aspects of rubber science, and was elected to the Rubber Science Hall of Fame in 1987, upon his death. He made careful studies of the crystallization of natural rubber. He used rheo-optical methods to study the local anisotropy of stretched rubber. He was a brilliant experimentalist and made seminal measurements of deformation in many different geometries. Treloar was

Fig. 5.3 L.R.G. Treloar
(Rubber Science Hall of
Fame, by permission)

**LESLIE RONALD GEORGE TRELOAR
(1906-1985)**

never satisfied to just make measurements; he became an expert in all theoretical aspects of rubber science.

While the Gaussian theory of rubber elasticity worked reasonably well for modest deformations, the finite extensibility of the chains becomes important for large strains. Treloar introduced the inverse Langevin function as a way to account for the limited extensibility. He also used numerical calculations to address the issue of the entropy of a rubber network. While crystallization intervenes in many cases, non-crystallizing rubbers have been shown experimentally to conform well to the Treloar approach. Treloar also encouraged Saunders and Rivlin to investigate appropriate equations of state for deformed isotropic networks. Ideal rubber theory assumes that the force of retraction is entirely due to the entropic contribution. Actual rubber does have an energetic contribution, and Treloar contributed both to the experimental measurement of this effect and to the theoretical understanding.

Actual polymer solutions and gels require both an understanding of the thermodynamics of mixing and the contribution of deformation to the chemical potential. Treloar and Gee made the seminal measurements on swollen rubbers, and Treloar explained the results for all the many types of deformations.

L.R.G. Treloar concluded his career at the University of Manchester Institute of Science and Technology, in the warm company of Sir Geoffrey Gee and Sir Geoffrey Allen.

George Stafford Whitby (1887-1972)

G. Stafford Whitby was a towering figure in rubber science from the earliest days of the 20th century to his death in 1972. He was very precocious and entered The Royal College of Science in London at 16. Upon graduation he continued in the laboratory of Sir William Tilden. In 1910 he was sent to Sumatra to study the production of natural rubber with the Societe Financiere des Caoutchoucs. He learned the rubber business from the ground up and made major contributions to the understanding of the chemical behavior of the raw latex (Fig. 5.4).

In 1917 he enrolled in graduate studies at McGill University in Montreal. He received his Ph.D. in organic chemistry in 1920. He remained at McGill and was promoted to Full Professor in 1923. His research continued in the area of rubber science and he developed successful methods of vulcanization, especially the use of accelerators. He obtained nine patents over the next ten years. Another milestone of his years at McGill was the publication of his monograph, *Plantation Rubber and the Testing of Rubber* [5]. It should be noted that in 1920 Whitby was discussing natural rubber in fully molecular and modern terms.

In 1929 Whitby assumed the post of Director of the Chemistry Division of the National Research Council of Canada Laboratories. In this administrative post, he continued his vigorous studies of polymerization. With the approach of war in Europe, he returned to his homeland as Director of the Chemical Research

Fig. 5.4 G. Stafford Whitby
(Rubber Science Hall of
Fame, by permission)

GEORGE STAFFORD WHITBY
(1887–1972)

Laboratories of the Department of Scientific and Industrial Research at Teddington, England. However, in 1942 he returned to North America as Professor of Rubber Chemistry at the University of Akron. His laboratory was immediately selected as one of the sites for the work of the Office of Rubber Reserve.

Whitby vigorously pursued research at Akron and produced another classic monograph, *Synthetic Rubber* [6]. While he "retired" in 1955, he continued to follow his many historical and philosophical interests. An admiring sketch in *Rubber World* in 1965 referred to him as a "Man for all seasons."

Roelof Houwink (1897-1988)

Like Herman Mark, Roelof Houwink was one of the featured speakers at the 1935 Faraday Discussion on Polymerization [9]. He represented the N.V. Phillips Gloeilampenfabricken at Eindhoven, Holland. He was still a major figure in materials science in 1953. He included polymers in all his work, but stressed the need to adopt a very broad paradigm in materials science. This advice is still very good! He communicated with all the leading workers in polymer science and convinced them to contribute chapters to many books that took a comprehensive approach to the structure and mechanical properties of polymers. One of his most famous books appeared entirely under his own name in 1937: *Elasticity, Plasticity and Structure of Matter* [10]. The second edition of this book closes the current period of interest (1953).

Houwink was born in 1897 in Holland and was educated at the University of Delft. He joined Phillips in 1925 and progressed up the ranks. In 1939 he was appointed the Director-General of the Rubber Institute at Delft. He served in this role until 1956. He was tireless in his promotion of a deep understanding of actual industrial chemicals, not just "ideal" materials. He studied a very wide range of substances, from inorganic glasses to asphalts. While his views of the glassy state would now be considered archaic, they had not been rejected in the period from 1935-1953. He was part of the school that viewed liquids as highly defective solids with "associated solid-like" regions in a low viscosity solvent. This "heterogeneous model" was very common for workers that studied primarily solids. Houwink was also fond of empirical relationships for measurable properties. The Mark-Houwink theory of the intrinsic viscosity contains a variable exponent. Many other rheological equations displayed this mathematical convenience. The more general use of power-law relations did not appear until "the French Revolution" in polymer science (1974-1995).

Walter H. Stockmayer (1914-2004)

Walter Stockmayer made beautiful "polymermusic" for more than 60 years. He was a Renaissance man with degrees from both MIT and Oxford, as a Rhodes Scholar. After his Ph.D. in 1940 he worked on war projects at Columbia before returning to MIT in 1945. In 1952 he was promoted to Full Professor and was elected to the National Academy of Sciences in 1956. He received the National Medal of Science in 1987 (Fig. 5.5).

Fig. 5.5 Walter Stockmayer (Dartmouth College, by permission)

He was a master of classical physical chemistry and specialized in the structure and dynamics of polymeric systems. He was both a good theorist and a careful experimentalist. He studied primarily solutions and gels. His comprehensive knowledge of polymer science led to many industrial consultantships, and he was welcomed internationally with both awards and foreign memberships. He was one of the founding editors of the journal *Macromolecules*.

John Douglass Ferry (1912-2002)

John D. Ferry is known as the leading figure in the history of polymer science on the subject of viscoelasticity. He graduated from Stanford University at the age of 19, as noted above. For his doctoral work with George Parks he studied the properties of polyisobutylene as a function of temperature. He found the glass transition temperature and characterized the viscoelastic properties (Fig. 5.6).

Ferry went to Harvard University in 1937 and worked there in a variety of posts, including as a Junior Fellow, until he joined the University of Wisconsin in 1946. He was promoted to Full Professor in 1947! His extensive measurements of the temperature dependence of the dynamic mechanical properties of polymers led to the concept of "reduced variables" in rheology. His demonstration that time-temperature superposition applied to many systems is the basis for the rational description of polymer rheology. He measured the dynamic response over a very wide range of frequency. One of the fruits of this work is the Williams-Landel-Ferry (WLF) equation for time-temperature shift factors.

Ferry also made seminal measurements on dilute solutions, using a precision shear wave apparatus. The Rouse-Zimm theory of chain dynamics was explored in great detail.

When the molecular chain length is long enough, polymeric systems display a rubbery plateau in the creep compliance. Ferry and his collaborators made shear creep measurements on a wide range of polymers. The asymptotic behavior of the shear viscosity was established.

The publication of his monograph, *Viscoelastic Properties of Polymers*, in 1961 provided a solid basis for all future work, and a key guide to all previous work [7]. It ranks with Flory's *Principles of Polymer Chemistry* as an essential record of the paradigm of polymer science.

Ferry's leadership role in polymer rheology was recognized early and he received the Bingham Medal of the Society of Rheology in 1953. He was elected to the National Academy of Sciences in 1959. He was also elected to the National Academy of Engineering and the American Academy of Arts and Sciences. He was inducted into the Rubber Science Hall of Fame upon his death in 2002. John D. Ferry was, in addition to being one of the intellectual leaders of polymer science, the heart and soul of the community of polymer scientists. Everyone who knew him loved him, and he made every meeting better by his presence.

Fig. 5.6 John D. Ferry (NAS biography, by permission)

Bruno Hasbrouck Zimm (1920-2005)

Bruno Zimm made a major impact on polymer science, both as an experimentalist and as a theorist. He learned the importance of statistical mechanics while earning his B.S and Ph.D. at Columbia University under the noted theorist, Joseph E. Mayer. Two other collaborators at Columbia were Paul Doty and Victor K. LaMer. In 1944 he moved to the Polytechnic Institute of Brooklyn where polymer research was in full swing. It was during this period that he teamed with his old friend Paul Doty and new friend Richard Stein to publish a light scattering classic [8]. Finally, in 1946 he was appointed to the University of California at Berkeley. During this period he developed the use of light scattering as a powerful tool in the study of the thermodynamics and structure of polymers in solution. After promotion to Associate Professor in 1950, he spent two years at Harvard. In 1951 Zimm accepted a position at the General Electric Laboratories. It was here that complete freedom allowed him to produce his seminal studies of the molecular dynamics of polymer chains in solution [12]. He also derived the so-called Zimm-Bragg theory of the helix-coil transition for chains in solution. He was elected to the National Academy of Sciences in 1958 (Fig. 5.7).

Fig. 5.7 Bruno Zimm (NAS biography, by permission)

Fig. 5.8 NAS Award winners in 1981. Author just to the right of Bruno Zimm. (Personal copy)

Zimm finished his long career at the University of California at San Diego, where he carried out exceedingly delicate measurements of biopolymers such as DNA. I remember fondly the night he received the National Academy of Sciences Award in the Chemical Sciences in 1981. I was also on the stage receiving the National Academy of Sciences Award for Initiatives in Research (Fig. 5.8).

Paul M. Doty (1920-2011)

Paul Doty was one of the brightest stars in the chemistry world during the period 1941-1957. He received his Ph.D. from Columbia University in 1944, but he had been doing research and teaching at Brooklyn Poly during the war since 1943. His thesis adviser at Columbia was the noted theorist Joseph Mayer. Doty was never intimidated by anything! At Brooklyn Poly he collaborated with Herman Mark, Bruno Zimm and Turner Alfrey. After the war, he went to Cambridge to work with Rideal. He developed his lifelong love of biopolymers during this stage. After a year at Notre Dame, he was appointed to Harvard University for the rest of his career. The world renowned program in molecular biology at Harvard owes its existence and excellence to Paul Doty. His election to the National Academy of Sciences in 1957 surprised no one (Fig. 5.9).

Fig. 5.9 Paul M. Doty
(Chemical Heritage
Foundation, by permission)

Richard S. Stein (1925-

Richard S. Stein was a true son of Brooklyn, New York. He graduated from the
Brooklyn Technical High School in 1942. He continued his education at Brooklyn
Polytechnic Institute. He was a very precocious undergraduate and published
seminal papers with Bruno Zimm and Paul Doty. After such a foundational education, what could follow it? Dick went to Princeton University and worked with
Arthur Tobolsky. After receiving his Ph.D. in 1948, he went to Cambridge University as an NRC Postdoctoral Fellow. He returned as a postdoctoral fellow to
Princeton. At this point Stein had learned perhaps more than anyone at that stage
of his career. He knew a large number of physical experimental techniques,
including X-ray and light scattering. He was thoroughly familiar with microscopic
techniques in both visible and infrared light. He knew the mechanical properties
of materials, including polymers. His subsequent career reflects this panoramic
background (Fig. 5.10).

In 1950, the University of Massachusetts at Amherst made one if it's most
intelligent decisions: it hired Richard S. Stein as a Chemistry Professor. Over the
next 63 years, Dick built the Polymer Research Institute into the premier polymer
program in America. While he had help from a good group of colleagues, Richard
S. Stein was the heart, soul and purse of UMass. He kept the research level at the
highest standard. He raised huge amounts of money for staff, equipment and
buildings. And he was universally recognized as a true polymer scientist of the
highest caliber.

Fig. 5.10 Richard S. Stein
(Who was who at UMass,
by permission)

William O. Baker (1915-2005)

Bill Baker rose from a bright graduate of Princeton to one of the most revered men in American science and industry. He retired as Chairman of the Board of AT&T Bell Laboratories in 1980. He worked with Charles P. Smyth at Princeton on the dielectric properties of organic crystals. He joined Bell Laboratories in 1939, after he received his Ph.D. in physical chemistry, and quickly applied his deep knowledge of dielectrics to problems in polymer science. His "solid state" perspective on polymers helped Bell Labs to become the pre-eminent research center in the world in condensed matter materials science.

While Bell Labs was not a "producer" of polymers, AT&T was one of the largest users in the world. The basic approach was to know more than anyone about the materials that were used for telecommunications. The British company ICI synthesized crystalline polyethylene, but Bell Labs discovered how to properly process this material and what microscopic parameters were important, such as the molecular weight distribution. DuPont synthesized nylon, but Bell Labs discovered how to make it useful as a stable material for long-term use. Cellulose polymers were an old technology, but Bell Labs discovered new ways to process it that achieved remarkable new properties and uses. Baker was the genius behind the philosophy and much of the early work. He wanted to know the actual truth about materials and reached out to the very best scientists. Peter Debye was a regular visitor to Bell Labs during the Baker era.

When the federal government established the Rubber Reserve Corporation, they recruited Bill Baker for the core team. He was present at the famous meeting in Akron in December 1942 where the rubber companies were nationalized. The key benefit of this arrangement was that everyone in the game shared information on a regular basis. For example, the program for the April 9, 1945 meeting included F.T. Wall, the convener from Illinois, W.O. Baker from Bell Labs, Peter Debye from Cornell, R.H. Ewart from U.S. Rubber, P.J. Flory from Goodyear, B.L. Johnson from Firestone, and E.J. Meehan from Minnesota. These teams created close bonds and the post-war leadership in polymer science was well known to each other from work on the rubber project. Most of the meetings were held in Akron, and Baker rode a lot of trains.

Fig. 5.11 William O. Baker (left) and Warren Mason discussing the viscoelastic properties of polymer solutions. (www.williamobaker.org)

By 1949 Bill Baker was the Head of the High Polymer Research and Development Department. Another member of this department, Field H. Winslow (1916-2009), had joined Bell Labs in 1945, after a career on the Manhattan project. Winslow went on to become the Editor of the journal *Macromolecules*. While William O. Baker rapidly rose to be Vice President for Research in 1955, he never forgot his roots in polymer science, and helped it to grow into the premier research laboratory in the area in the world. He would frequently drop by my laboratory and discuss my research on the structure and dynamics of liquid and glassy polymers in detail. Not only did he know what I was doing, he had thought deeply about it (Fig. 5.11).

References

1. Ott E (1943) Cellulose and cellulose derivatives. High Polymers, Volume V, Interscience Publishers, New York
2. Burk RE, Grummit O (1943) The chemistry of large molecules. Interscience Publishers, New York
3. Huggins ML (1936) Hydrogen bridges in organic compounds. J Org Chem 1:407–456
4. Treloar LRG (1949) The physics of rubber elasticity. Oxford at the Clarendon Press, Oxford
5. Whitby GS (1920) Plantation rubber and the testing of rubber. Longmans, Green and Co., London

6. Houwink R (1936) The strength and modulus of elasticity of some amorphous materials, related to their internal structure. Trans Faraday Soc 32:122–131
7. Ferry JD (1961) Viscoelastic properties of polymers. John Wiley and Sons, New York
8. Zimm BH, Stein RS, Doty P (1945) Classical theory of light scattering from solutions—a review. Polym Bull 1:90–119

References

Chapter 6
Hermann Staudinger and the Nobel Prize for Chemistry 1953

A Prehistory of Polymer Science ended with Hermann Staudinger well ensconced at Freiburg and basking in the success of the Faraday Discussion on Polymerization [1]. But better days were not ahead. Most of his best collaborators were either going or gone. The conditions in Germany were not going to get better and all scientists were expected to serve the German war machine. Staudinger's reward for his "service" in the War was that his Institute was badly damaged. After a huge effort to rebuild his Institute, he retired in 1951. His 50 years of labor in polymer science was recognized with the Nobel Prize in Chemistry for 1953 (Fig. 6.1).

The citation read "for his discoveries in the field of macromolecular chemistry." He was very pleased and prepared an extensive acceptance lecture. He followed this with a more extensive scientific autobiography in 1961, *Arbeitserinnerungen* [2]. Eventually, after his death, this book was translated into English and issued as *From Organic Chemistry to Macromolecules: A Scientific Autobiography based on my Original Papers* [3].

It is instructive to analyze the Nobel acceptance speech, to get a picture of Staudinger's view of his own work in 1953. Even in 1953, Staudinger was still under the impression that macromolecular chemistry was just getting started, and that the award of the Nobel Prize to him would greatly assist the progress of the field. He had virtually no sense that the field had passed him by, or who it was that was currently leading the community of polymer scientists.

Ever the demarcator, he decreed that macromolecules had at least 1000 atoms covalently bonded into a particular structure. He stressed that true macromolecules had "properties" that differed from "low molecular compounds." He could never quite give up the hope that Macromolecular Science would be a field of its own with its own distinctive laws and principles, like Colloid Science.

No one could accuse Hermann Staudinger of being lazy. He was highly prolific and published more than 400 papers between 1926, when he moved to Freiburg, and 1953.

Early chemistry struggled with the existence of both "natural" products and purely "synthetic" substances that were not found free in Nature. "Health food" stores still trade on this distinction. Staudinger also could not quite abandon the notion that natural polymers were unique and not capable of being synthesized.

G. Patterson, *Polymer Science from 1935–1953*, SpringerBriefs in History of Chemistry, DOI: 10.1007/978-3-662-43536-6_6, © The Author(s) 2014

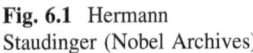

Fig. 6.1 Hermann
Staudinger (Nobel Archives)

There is substantial evidence that synthetic attempts to produce "natural rubber" yielded a product that was not identical with the material isolated from caoutchouc. There is also evidence that there are as many "polyethylenes" as there are attempts to synthesize it! There are times when biological systems exercise exquisite control over both the local and global structure of macromolecules. The key to creating polyisoprene that performed as well as natural rubber was synthetic control using the right catalyst and the best conditions. That level of control was visible in 1953, but not to Staudinger.

Staudinger had performed extensive studies of the polymerization of styrene. It might be supposed that by 1953 he would be able to summarize the state of the art in exquisite detail. He obfuscated the initiation of pure styrene as a "magical activation." He might be forgiven this sin since the details of styrene initiation are still under discussion. He claimed that the "terminated" chain still had a terminal unsaturation. After the careful work of Flory on radical polymerization, this is not forgivable.

Hermann Staudinger was aware of the many challenges associated with the analytical chemistry of macromolecules. Since many macromolecules do not crystallize at all, the melting point cannot be used as a qualitative test. It can also be difficult to prepare homogeneous solutions of many macromolecules. However, Staudinger was more at home with the likes of Wolfgang Ostwald than with Svedberg, Mark and Meyer. Even his collaborator Signer kept his distance after moving to Bern. Staudinger retained many of the worst aspects of W. Ostwald's "Kolloid Science."

At other times Staudinger led the way by utilizing controlled chemical modification of macromolecules to reveal details of their structure. One of his most famous examples was the full hydrogenation of natural rubber to yield a very high molecular weight macromolecule. He was even aware of the significance for biological systems of chemically stable macromolecules that could retain their structure during biological processes.

In spite of the detailed work on both the osmotic pressure and intrinsic viscosity of polymer solutions by Gee and Flory, Staudinger continued to use his "viscosity law." He even still viewed the molecules as highly extended. While Hermann Staudinger should be praised for his industry and persistence in promoting the macromolecular paradigm, he ossified in his own thinking in the late 1930s and did not contribute to the consolidation of the paradigm in the period 1935-1953.

References

1. Patterson G (2012) A prehistory of polymer science. Springer, New York
2. Staudinger H (1961) Arbeitserinnerungen. Verlag GmbH, Heidelberg
3. Staudinger H (1970) From organic chemistry to macromolecules: a scientific autobiography based on my original papers. Mark, HF trans. Wiley-Interscience, New York

Chapter 7
Concluding Reflections and Further Thoughts

The period from 1935-1953 was a very profitable and dynamic era in the history of polymer science. The worldwide community that coalesced in 1935 was very active in identifying the key phenomena and forging satisfying qualitative and quantitative theories. The state of the community in 1953 will be surveyed from several perspectives.

Just in terms of sheer numbers, the community grew strongly in this period from a few hundred to a few thousand. They occupied positions within industry, government laboratories, independent laboratories and academic institutions. In the United States, there were now many companies that produced polymers: DuPont, Harshaw Chemical Company, Bakelite Corporation, Kodak, General Electric, Celanese Corporation, Goodyear Tire and Rubber, Rohm and Haas, Dow Corning Corporation, Standard Oil Company, Hercules Powder Company, Dow Chemical, American Cyanamid, Bell Telephone Laboratories, Firestone Tire and Rubber, General Tire and Rubber, B.F. Goodrich, U.S. Rubber, Phillips Petroleum, Proctor and Gamble, Shell Chemical Corporation, Union Carbide, and others.

The United States Bureau of Standards had a strong and growing polymer research group and the Mellon Institute maintained serious research in polymers. Academic programs at the Brooklyn Polytechnic Institute, University of Akron, Western Reserve University, The University of Massachusetts, the University of Illinois, Princeton University, Cornell University, and others were thriving.

The situation in Europe was also strong and growing. British companies such as Imperial Chemical Industries (ICI) and universities such as Manchester were centers of polymer research. The National Physical Laboratories at Teddington was poised to produce several decades of leading research. In Holland, both companies like Phillips and research organizations like the Netherlands Organization for Applied Scientific Research (TNO) were centers of polymer activity. In France, companies like Michelin and Rhone-Poulenc and universities like Strasbourg had strong polymer programs. Montecatini in Italy was a strong international competitor. And Germany had recovered enough that Bayer at Leverkeusen and the Max Plank Institute for Polymerforschung were leading the way. Everything was ripe for a long period of sustained progress.

G. Patterson, *Polymer Science from 1935–1953*, SpringerBriefs in History of Chemistry, 79
DOI: 10.1007/978-3-662-43536-6_7, © The Author(s) 2014

The international field of polymer science now had a strong collection of true scientific leaders to guide the growth and application of knowledge. Herman Mark was the leading person in polymer science throughout the world. He was looked up to by virtually all polymer scientists. And he tried to promote both polymer science and other polymer scientists. His success was spectacular.

With many workers now committed to the study of polymers, groups of scientists were formed in many places. Some of the groups were academically based. In addition to Brooklyn Poly, the University of Akron, Case Western, the University of Massachusets, Princeton University, and many others will be examined in future books. Some of the groups were in government laboratories. Especially strong efforts at the National Bureau of Standards, at the National Physical Laboratory in England, and at the Max Planck Institute in Germany were active in this period. And there were also independent research laboratories such as the Mellon Institute for Industrial Research.

Most of the individual leaders introduced in the present volume continued into the next period. They frequently met together at national meetings of the American Chemical Society, the American Physical Society and at Gordon Research Conferences. They served together on the boards of journals such as the Journal of Polymer Science and Macromolecules. All the externalities of a thriving research community emerged in the period from 1953-1974. Even money for research was regularized during this period.

There were also new young leaders that achieved prominence during the next period. They came from a very wide range of places. From the industrial world of Bell Laboratories, people like Frank Bovey and Field Winslow emerged. From the Bureau of Standards, leaders like John Hoffman hired an exceptionally strong staff. While Stafford Whitby was the early soul of the University of Akron polymer effort, Frank Kelly built the premier institution in the City of Akron during this period. He recruited a great staff and things have only gotten better. In France leaders like Henri Benoit at Strasbourg and Champetier at the Ecole Superieure de Physique et de Chimie Industrielle de la Ville de Paris built strong programs. In Germany Erhard Fischer gathered a vigorous group in Mainz. In Japan, Kurata, Yamakawa and Fujita were the core of a great effort at Kyoto.

Much of the early polymer science was carried out with primitive tools like the Ostwald viscometer. The period 1953-1974 saw the appearance of truly sophisticated instrumentation that proved useful in polymer science. Nuclear magnetic resonance spectroscopy was improved from a curiosity of the physics laboratory to a powerful instrumental method in chemistry at Bell Laboratories by William Slichter, David McCall, Dean Douglas and Frank Bovey. The primitive calculations of previous eras were revolutionized by the perfection of the digital computer. Mechanical spectroscopy was extended into the hypersonic range by William O. Baker and many others. Dielectric spectroscopy was also extended into the gigahertz range by Robert H. Cole at Brown and widely used at NPL and Bell Labs. While light scattering was known and practiced before 1953, the invention of the laser revolutionized its practice for polymer science. Precision calorimeters allowed polymer science to become truly quantitative with regard to thermodynamic properties. As the maturity of the field increased, the ability to use the best equipment to ask and answer questions in polymer science improved.

Once the level of precision was high enough and the certainty of the phenomenology was assured, theorists discovered the wealth of problems involving polymers. At Mellon Institute Edward Cassasa applied thermodynamics and statistical mechanics to polymers. Marshall Fixman joined Kirkwood at Yale in a similar effort. Stuart Rice at Chicago joined Fuoss in examining polyelectrolytes. By 1974 theoretical polymer science was an exciting field, and the best was yet to come.

The biggest problem facing polymer science in 1953 was the sustainability of the funding for future work. With the demise of the Rubber Reserve Corporation, one major source disappeared. The National Science Foundation established a small effort in polymer science. But, industry itself discovered that investment in polymer research could be profitable. Major expansions in research throughout the worldwide polymer industry led to one new discovery after another. And the military establishments decided they could not do without unique polymers, both in the United States and in the Soviet Union. While individual workers needed to be entrepreneurial, there was money to be found somewhere.

About the Author

Gary Patterson is Professor of Chemical Physics and Polymer Science at Carnegie Mellon University in Pittsburgh, Pennsylvania. He is the Chief Bibliophile of the Bolton Society at the Chemical Heritage Foundation in Philadelphia, Pennsylvania. He was educated at Harvey Mudd College (B.S. Chemistry, 1968) and Stanford University (Ph.D. Physical Chemistry, 1972). His thesis advisor was Paul Flory, who received the Nobel Prize in Chemistry for his work in Polymer Science in 1974. He was a Member of Technical Staff in the Chemical Physics Department at AT&T Bell Laboratories from 1972–1984, when he joined the Chemistry Department at Carnegie Mellon University. He was the Charles Price Fellow at the Chemical Heritage Foundation in 2004–2005. He is a Fellow of the Royal Society of Chemistry and the American Physical Society. He received the National Academy of Sciences Award for Initiatives in Research in 1981 for his research in Polymer Science. Gary is the Chair-elect of the History Division of the American Chemical Society.

G. Patterson, *Polymer Science from 1935–1953*, SpringerBriefs in History of Chemistry, DOI: 10.1007/978-3-662-43536-6, © The Author(s) 2014

About the Author

Author Index

A

Adams, Roger (1889-1971), 2, 5, 51
Alfrey, Turner (1918-1981), 2, 41, 69
Allen, Geoffrey (1928-), 10, 63
Astbury, W. T. (1898-1961), 39

B

Baker, William O. (1915-2005), 3, 71, 72, 80
Berthelot, Marcellin (1827-1907), 29
Bolton, Elmer K. (1886-1968), 5, 51
Boundy, Ray H. (1902-1992), 34
Boyer, Raymond F. (1910-1993), 34
Bueche, Arthur M. (1920-1981), 18
Bunn, C. W. (1916-1976), 32
Burk, Robert E. (1901-1978), 28

C

Carothers, Wallace Hume (1896-1937), 1, 2, 5, 6, 29, 51
Conant, James Bryant (1893-1978), 5, 29, 31, 51

D

Daniels, Farrington (1889-1972), 51
Debye, Peter (1884-1966), 2, 3, 14, 17, 71
Doty, Paul M. (1920-2011), 3, 68-70

E

Eirich, Frederick (1905-2005), 25

F

Ferry, John Douglass (1912-2002), 46, 66, 67
Flory, Paul John (1910-1985), 1, 2, 6, 20, 71

Fox, Thomas G (1921-1977), 2, 13, 16-18
Fuoss, Raymond Matthew (1905-1987), 61

G

Gee, Geoffrey (1910-1996), 10, 11, 62, 63
Guth, Eugene (1905-1990), 10, 26

H

Haas, Otto (1872-1960), 32, 34
Houwink, Roelof (1897-1988), 3, 27, 47, 64
Huggins, Maurice Loyal (1897-1981), 26, 47, 60

I

Ipatieff, V. N. (1867-1952), 29, 30

K

Katz, J. R. (1880-1938), 38
Kirkwood, John Gamble (1907-1959), 18, 42
Kraemer, Elmer O. (1898-1943), 27, 28
Krause, Charles A. (1876-1967), 61
Krigbaum, William R. (1922-1991), 2, 14, 20
Kuhn, Werner (1899-1963), 2, 3, 7, 15, 25

L

Lebedev, S. V. (1874-1934), 29

M

Mandelkern, Leo (1922-2006), 2, 19, 20
Mark, Herman Francis (1895-1992), 2, 3, 5, 10, 20, 23-25, 38, 45, 49, 53, 54, 64, 65, 69, 80

G. Patterson, *Polymer Science from 1935–1953*, SpringerBriefs in History of Chemistry,   85
DOI: 10.1007/978-3-662-43536-6, © The Author(s) 2014

Subject Index

G. Patterson, *Polymer Science from 1935–1953*, SpringerBriefs in History of Chemistry,   87
DOI: 10.1007/978-3-662-43536-6, © The Author(s) 2014